A Genetically Modified Future?

ISSUES

Volume 138

Series Editor

Lisa Firth

 **Independence**

Educational Publishers
Cambridge

First published by Independence
PO Box 295
Cambridge CB1 3XP
England

© Independence 2007

British Library Cataloguing in Publication Data
A Genetically Modified Future? – (Issues Series)
I. Firth, Lisa II. Series
363.1'92

ISBN 978 1 86168 390 8

Printed in Great Britain
MWL Print Group Ltd

Cover
The illustration on the front cover is by
Don Hatcher.

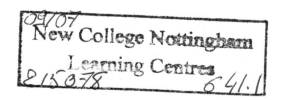

CONTENTS

Chapter One: GM Trends

Chapter Two: The GM Debate

Introduction

A Genetically Modified Future? is the one hundred and thirty-eighth volume in the **Issues** series. The aim of this series is to offer up-to-date information about important issues in our world.

A Genetically Modified Future? looks at GM trends and the debate about genetic modification.

The information comes from a wide variety of sources and includes:
Government reports and statistics
Newspaper reports and features
Magazine articles and surveys
Website material
Literature from lobby groups
and charitable organisations.

It is hoped that, as you read about the many aspects of the issues explored in this book, you will critically evaluate the information presented. It is important that you decide whether you are being presented with facts or opinions. Does the writer give a biased or an unbiased report? If an opinion is being expressed, do you agree with the writer?

A Genetically Modified Future? offers a useful starting-point for those who need convenient access to information about the many issues involved. However, it is only a starting-point. Following each article is a URL to the relevant organisation's website, which you may wish to visit for further information.

* * * * *

What is GM?

How is it done? Where is it done? Does it have to be done in a lab?

'GM' stands for genetic modification. A gene is an instruction and each of our cells contains tens of thousands of these instructions. In humans, these instructions work together to determine everything from our eye colour to our risk of heart disease. The reason we all have slightly different characteristics, even from our brothers and sisters, is that before we are born our parents' genes get shuffled about at random.

GM allows chosen individual genes to be transferred from one organism into another, including genes between non-related species. Such methods can be used to create GM crop plants

Exactly the same principles apply to plants. If you're a gardener, you might save seed from a favourite plant, hoping you'll get another plant exactly the same. But, because genes get shuffled about, you might get something that looks rather different. It's still the same kind of plant but it's bigger or smaller or a slightly different colour.

For thousands of years farmers have selected plants with the characteristics they want, such as extra seeds in a pod or the ability to survive in the cold. By crossing the best plants they hoped to produce even better varieties. But this approach is a bit like playing a fruit machine: you hit the jackpot only very occasionally. So since the 1950s, plant scientists have lent a hand.

Deliberately exposing seeds to radiation, for instance, increases the chance that one of them might produce a more useful plant. The barley variety Golden Promise was produced in this way and has been growing in Britain for 30 years.

Unlike these earlier methods, GM techniques allow specific genes to be copied into a plant. Because the scientists know a great deal about the genes they're working with, it's easier for them to 'track' the genes, understand their effects and eliminate unwanted side effects long before the plants are used in field trials or grown commercially.

GM allows chosen individual genes to be transferred from one organism into another, including genes between non-related species. Such methods can be used to create GM crop plants. The technology is also sometimes called 'modern biotechnology', 'gene technology', 'recombinant DNA technology' or 'genetic engineering'.

The actual transfer of genes into the selected organism (a plant for example) always takes place in a laboratory under carefully controlled conditions. Genetically modified plants can later be trialled in a special glasshouse or in fields under regulatory oversight, before being grown commercially, using systems long in use for testing and evaluating new plant varieties.

⇨ Information from CropGen – making the case for crop bio-technology. Visit www.cropgen.org for more information.

© CropGen

Genomics in the UK

Information from the Economic and Social Research Council

This article examines genomics from a social science perspective. It defines the terms and provides a brief history and summary of genomic science. It then focuses on GM crops and reviews the politics of genomics in the UK and provides information and statistics on the current situation.

What is genomics and genetic modification?

Contained in every cell of every organism are chromosomes, genes and deoxyribonucleic acid (DNA) and these together are said to constitute the 'genetic information' or 'genetic material' that carries the instructions for all the characteristics that an organism inherits. An organism's genome can be defined as all this genetic material considered together. Genomics is the science of genomes – more specifically, their sequencing, mapping, analysis, study and manipulation. In theory it is thought that any living organism can have its genome modified – including humans. Genomics should not be confused with the closely-related topic of genetics, which is the scientific study of heredity.

Genetic modification (GM) refers to moving genetic material from the cells of one organism to another, be they related or unrelated.

History of genetic modification

Contemporary GM techniques are based on scientific discoveries made in the 1950s. Research into molecular biology and genetics in the 1970s resulted in the first GM plants being bred during the early 1980s. The first commercial crops were grown on a large scale in 1996.

How does GM work?

For the genetic modification of crops, a plant that has the desired characteristic is first identified. The specific gene that produces this characteristic is located and cut out of the plant's DNA. A piece of bacterial

DNA, called a plasmid, is joined to the gene to act as the carrier that gets the gene into the cells of the plant being modified.

Research into molecular biology and genetics in the 1970s resulted in the first GM plants being bred during the early 1980s. The first commercial crops were grown on a large scale in 1996

Furthermore, a 'promoter' is also included with the combined gene and plasmid. This helps make sure the gene works properly when it is put into the plant being modified. This gene package is then inserted into a bacterium, which is allowed to reproduce to create many copies of the gene package.

Finally, the gene packages are transferred into the plant being modified. The plant tissue that has taken up the gene packages is then grown into full-size GM plants.

The global GM crop picture

The most immediate issue surrounding genetic modification surrounds GM crops. Food crops are modified for a variety of reasons, including to produce a greater yield of crop or to be more resilient to pesticides. In

2005 21 countries grew genetically modified crops compared with 17 in 2004. The table 'Global area of GM crops in 2005' shows the 21 GM crop-producing countries and their area of GM crop cultivation.

In 2005, GM crops grown worldwide covered 90 million hectares which is three times the land area of Britain, a 10 per cent increase on 2004. The total worldwide area of GM crop cultivation is growing rapidly. However, GM crops still make up less than five per cent of the total global area of crop growing. By land area, the vast majority (99 per cent) are grown in the US, Argentina, Canada and China. The US accounts for around two-thirds of the world total.

In 2005 the global market value for GM crops was estimated at $5.25 billion (£2.69 billion).

There are currently no GM crops being grown in the UK. In recent years GM crops have been grown for research and development purposes

at a number of sites as part of the UK farm scale evaluation trials. No GM crops are expected to be grown commercially in the UK before 2009.

The politics of genomics

Public attitudes to genetic modification in the UK are complex and constantly shifting. A report by the University of Manchester suggests that overall public attitudes towards GM foods are still negative with 47 per cent of people feeling that GM crops should not be grown for commercial use and 29 per cent of people still undecided on the issue. See the graph 'Public attitudes towards GM crops in the UK'.

To gauge the UK public's opinion of GM the UK government organised a national public debate in June 2003. This debate revealed five key characteristics of the UK public's attitude to GM:

⇨ People are generally uneasy about genetic modification .

⇨ The more people engage with genetic modification issues the harder their attitudes and deeper their concerns become.

⇨ There is little support for the early commercialisation of genetic modification.

⇨ There is a widespread distrust of multinational corporations and the government in relation to genetically modified organisms (GMOs).

⇨ There is a broad desire for more knowledge of genetic modification and for more research to be carried out.

Regulation of genetic modification in the UK

Regulations governing genetic modification were developed because of concerns over the possible adverse effects on biodiversity and human health of the release of GMOs and due to the ethical implications of modifications to humans and higher mammals.

The release and marketing of GMOs in the EU is governed by European Directive 90/220/EEC. On a UK scale all applications are scrutinised on a case-by-case basis by the Advisory Committee on Releases to the Environment (ACRE), an independent scientific committee. Furthermore, all food in the EU that contains or consists of GMOs must be labelled accordingly.

⇨ The above information is reprinted with kind permission from the Economic and Social Research Council. For more information, please visit the ESRC website at www.esrc.ac.uk

© ESRC

Genomics in the UK

Global area of GM crops in 2005. 1 hectare = 2.47 acres

Country	Area (million hectares)	GM crops
USA	49.8	soybean, maize, cotton, canola, squash, papaya
Argentina	17.1	soybean, maize, cotton
Brazil	9.4	soybean, maize, canola
Canada	5.8	soybean
China	3.3	cotton
Paraguay	1.8	soybean
India	1.3	cotton
South Africa	0.5	maize, soybean, cotton
Uruguay	0.3	soybean, maize
Australia	0.3	cotton
Mexico	0.1	cotton, soybean
Romania	0.1	soybean
Philippines	0.1	maize
Spain	0.1	maize
Colombia	less than 0.05	soybean
Iran	less than 0.05	rice
Honduras	less than 0.05	maize
Portugal	less than 0.05	maize
Germany	less than 0.05	maize
France	less than 0.05	maize
Czech Republic	less than 0.05	maize

Source: James, C (2005), Global status of commercialised biotech/GM crops, International Service for the Acquisition of Agri-biotech Applications (ISAAA) pp6 (accessed 19 November 2006). Taken from the ESRC factsheet 'Genomics in the UK'.

Public attitudes towards GM crops in the UK

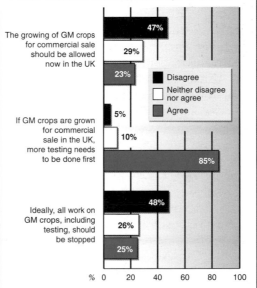

The growing of GM crops for commercial sale should be allowed now in the UK — Disagree 47%, Neither disagree nor agree 29%, Agree 23%

If GM crops are grown for commercial sale in the UK, more testing needs to be done first — Disagree 5%, Neither disagree nor agree 10%, Agree 85%

Ideally, all work on GM crops, including testing, should be stopped — Disagree 48%, Neither disagree nor agree 26%, Agree 25%

Rigby, D, Young, T, and Burton, M (2004), Consumer willingness to pay to reduce GMOs in food and increase the robustness of GM labelling, University of Manchester, pp58 [PDF 78] (accessed 30 November 2006). Taken from the ESRC factsheet 'Genomics in the UK'.

Glossary

Information from GM Nation?

Agro-chemicals
Chemicals for use in the agricultural industry.

Allergen
A substance which causes an immune response.

Antibiotic-resistant
The ability of a micro-organism to disable an antibiotic or prevent its transport into the cell.

Biofuel
A gaseous, liquid or solid fuel derived from a biological source, e.g. ethanol, rapeseed oil or fish liver oil.

Biotechnology
Any technological application that uses biological systems, living organisms, or derivatives thereof, to make or modify products or processes for specific use.

Cross-contamination
Unintended insertion of a gene or genes from one organism into another. It sometimes refers to the unintended presence of GM material in a non-GM crop or food.

Deliberate Release Directive
Directive 2001/18/EC represents EU legislation designed to protect health and the environment across the EU from any adverse effects that may be caused by the deliberate release into the environment of genetically modified organisms.

DNA
Abbreviation for deoxyribonucleic acid. DNA constitutes the genetic material of most known organisms and organelles, and usually is in the form of a double helix.

Gene
A piece of DNA code, an instruction to build a protein which then forms part of, or does work in, a body. Sometimes, a single gene determines an effect. But most processes that build and maintain bodies and plants involve many genes.

Genetic modification (GM)
The technology of altering the genetic material of an organism by the direct introduction of DNA.

Genetic modification may also sometimes be called 'modern biotechnology', 'gene technology', 'recombinant DNA technology' or 'genetic engineering'.

Genetically modified organism (GMO)
A genetically modified organism is one in which the genetic material has been altered by the direct introduction of DNA.

Herbicide tolerance
The ability of a plant to remain unaffected by the application of a herbicide.

Marker gene
A gene or short sequence of DNA that acts as a tag for another, closely linked, gene.

A genetically modified organism is one in which the genetic material has been altered

---

Patent
A form of legal protection giving monopoly rights over exploitation of an invention for 20 years.

Pharmaceutical (or 'pharmed') crops
Crops that are genetically modified for pharmaceutical purposes.

Protein
A complex molecule consisting of a particular sequence of chains of amino acids. Proteins are essential constituents of all living things.

Technology fee
Fee charged by owners of a patent for the use of their invention.

⇨ The above information is re-printed with kind permission from GM Nation? Visit www.gmnation.org for more information.

© GM Nation?

GM (Genetic Modification)

Answers to some frequently asked questions about GM crops

Current situation/policy

Q. What is the current situation with GM crops in the UK?

There are currently no GM crops being grown in the UK. They have previously been grown for research and development purposes at a number of sites. The main example of this was the Farm-Scale Evaluation (FSE) trials that ended in 2003. No GM crops are expected to be grown here commercially before 2009 at the earliest. For the foreseeable future, the only crops likely to be proposed for cultivation in the UK are commodity crops – e.g. oilseed rape – not horticultural products like fruits and vegetables.

There are currently no GM crops being grown in the UK

Q. What is the Government's policy on GM crops?

The Government set out its policy on the commercial growing of GM crops in a statement to Parliament in March 2004. This reflected a careful evaluation of all the available information including the reports from the GM public debate, science review and costs and benefits study. No other country has undertaken such a comprehensive assessment of the case for and against GM crops. Having weighed all the evidence, we have concluded that the only sensible approach is to assess each GM crop on an individual case-by-case basis.

The Government's top priority is to protect human health and the environment. Under European Union (EU) legislation each proposed release of a GM product is subject to a detailed risk assessment which involves careful scrutiny by independent scientists. An evaluation is made of all the risk factors that may arise, including possible toxic or allergenic effects and the likely consequences of any gene transfer. The Government takes a precautionary approach and we will only agree to the commercial cultivation of a GM crop if we are satisfied that it is safe. To promote transparency and greater public confidence in the way these decisions are made, this independent advice is published on the Defra website (www.defra.gov.uk/environment/gm/regulation/euconsent.htm) along with the Government's subsequent conclusions.

Q. Hasn't the Government already agreed to the commercial cultivation of GM crops?

In our GM policy statement we gave our specific view on three GM crops grown in the Farm-Scale Evaluation trials. We confirmed that we would not agree to commercial approval for the GM beet and spring oilseed rape involved, because the scientific evidence suggested that the herbicide use associated with these crops – although not the GM plants themselves – could have an adverse effect on the environment. However, we agreed in principle to the commercial cultivation of the GM maize grown in the trials, subject to certain conditions. In the event, Bayer CropScience subsequently announced that they would not in fact market this particular GM maize variety.

Q. When will approved GM crops be grown in the UK?

We do not expect any commercial cultivation of GM crops in the UK before 2009 at the earliest, but beyond that it is not possible to say when the first crop might be introduced here and what type of crop it might be.

Decisions on the commercial approval of GM crops are taken collectively at European Union (EU) level on a case-by-case basis. Various applications to grow GM crops are currently working their way through the EU system. Details on the applications and their progress is available on our website (www.defra.gov.uk/environment/gm/regulation/registers.htm). A small number of GM crop varieties have all the approvals necessary to be grown. However, these are maize varieties which are suitable for cultivation in Southern Europe and which are resistant to pests which are not present in the UK. Therefore, we do not expect them to be grown in the UK.

Q. What type of GM crops are being developed?

There are two main types of GM crops that are in commercial use around the world. These are either

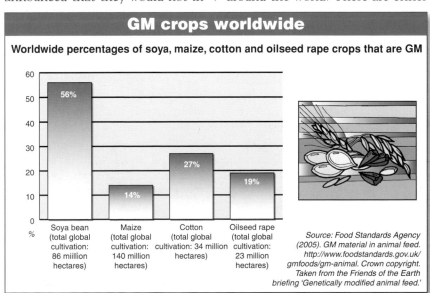

GM crops worldwide

Worldwide percentages of soya, maize, cotton and oilseed rape crops that are GM

Soya bean (total global cultivation: 86 million hectares)	Maize (total global cultivation: 140 million hectares)	Cotton (total global cultivation: 34 million hectares)	Oilseed rape (total global cultivation: 23 million hectares)
56%	14%	27%	19%

Source: Food Standards Agency (2005). GM material in animal feed. http://www.foodstandards.gov.uk/gmfoods/gm-animal. Crown copyright. Taken from the Friends of the Earth briefing 'Genetically modified animal feed.'

crops that have been developed to be resistant to certain crop pests, or crops that have been developed to be resistant to a particular herbicide (weed killer). These GM traits are being used in crops such as soya, maize, oilseed rape and cotton.

Safety of GM crops

Q. How is the safety of GM crops assessed?
Under EU legislation each proposed release of a GM product is subject to a detailed risk assessment which involves careful scrutiny by independent scientists. An evaluation is made of all the risk factors that may arise, including possible toxic or allergenic effects and the likely consequences of any gene transfer to other organisms. This takes account of relevant evidence from tests and trials as well as existing scientific knowledge. Ensuring safety is the Government's clear over-riding objective on this issue. We take a precautionary approach and will only agree to a GM crop release if we are satisfied that it is safe.

More information about how GM crops are regulated is available in the GM regulation section of Defra's website: www.defra.gov.uk/environment/gm/regulation/index.htm
Q. What about long-term effects?
The EU controls operate on a step-by-step basis. The initial development and testing of a new GM crop is done in the laboratory and it will only be released into the environment for further work when this is deemed sufficiently safe. The first release into the open will be at a small scale with precautionary measures taken to isolate the GM plants from the wider environment. It is only after these initial stages that approval might be given for a larger-scale field release, assuming that the accumulated evidence supports the view that it is safe to proceed. Overall, therefore, a GM crop is subject to an extended process of tests, trials and evaluation before it may be approved for commercial use. In addition, the controls require a GM crop to be monitored after it has been cleared for marketing, to check whether the assumptions made about safety are confirmed in practice.

Q. But isn't more long-term research needed to prove there is no risk?
Every activity carries some degree of risk and it is not possible for research to eliminate all conceivable risk, no matter how long-term it is. A judgement has to be made about the evidence needed to make a balanced decision. As noted above, GM crops are subject to a rigorous assessment procedure which will ensure that the potential risks are considered very carefully. If there is a need for specific research to be undertaken we would expect this to be identified as part of the evaluation process.

It is true that with current GM techniques there are potential uncertainties about how inserted genes may perform, and this means it is possible for unexpected effects to occur

---

Q. Isn't it the case that genetic modification is an unstable process and genes are inserted at random to make a GM plant – doesn't this makes unexpected effects likely?
It is true that with current GM techniques there are potential uncertainties about how inserted genes may perform, and this means it is possible for unexpected effects to occur. But this is also the case with conventional plant breeding. Both GM and conventional varieties undergo extensive testing to select those which have desirable traits and reject those which do not. If a newly-created GM plant has an unstable genetic structure this is likely to manifest itself early on, resulting in that plant not being developed any further. A GM crop will go through years of research and development and many generations of plant breeding before it is ready for possible commercial use. By that time, if the GM plant is growing in a stable and reliable fashion it may

reasonably be concluded that the novel genes have been inserted stably into the plant's genome.
Q. What about cross-pollination between GM and non-GM crops?
Cross-pollination is a normal process between sexually compatible plants and the fact that a GM crop may transfer genes to other plants does not mean that there will be an adverse effect on the environment. Before a GM crop can be approved for release the consequences of any potential cross-pollination are carefully evaluated as part of the statutory risk assessment process. In addition we recognise that GM cross-pollination could affect the economic interests of conventional and organic farmers.

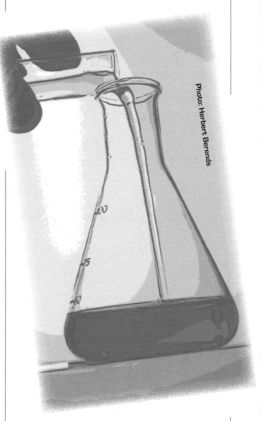

Photo: Herbert Berends

The Government's Advisory Committee on Releases to the Environment ACRE has said that if GM herbicide-tolerant crops transfer genes to other crops or wild relatives this poses a very low environmental risk. This is because the herbicide-tolerance genes are not likely to give the recipient plants any competitive advantage that would make them more invasive or 'weedy'. Each GM crop has to be considered on a case-by-case basis because they may differ in their potential impact.

Q. *But if anything does go wrong the impact will be irreversible.*
The effects of a GM crop are not necessarily irreversible. For example, in this country maize has no wild relatives and does not shed viable seed, so if a GM maize is grown the GM genes do not persist in the environment. Even for a crop like oilseed rape that does have wild relatives and feral populations and can release viable seed, the available evidence indicates that GM herbicide-tolerance genes are unlikely to persist. This is because if wild plants are cross-pollinated by oilseed rape crops the resulting hybrid is normally 'stunted' and unlikely to reproduce. In addition herbicide-tolerance genes do not confer any selective advantage to feral plants in unmanaged areas. Again, each GM crop, and the risk that it may pose, has to be considered on a case-by-case basis.

Under EU legislation each proposed release of a GM product is subject to a detailed risk assessment which involves careful scrutiny by independent scientists

The Farm-Scale Evaluation (FSE) trials
Q. *What did the trials investigate?*
The FSE trials measured the impact on farmland wildlife of the herbicide use associated with four GM herbicide-tolerant crops, as compared to the herbicide use with the equivalent conventional crops. They were not designed to look at the impact of the GM crop itself. The four crops are: spring-sown oilseed rape, maize and beet, and autumn-sown oilseed rape.
Q. *What did the results show?*
The results for the spring-sown crops (oilseed rape, beet and maize) were published on 16 October 2003. The Advisory Committee on Releases to the Environment were asked to

advise ministers on their implications and they published their advice on 13 January 2004.
ACRE advised that if the crops were to be grown and managed as in the FSEs:
⇨ there would be no adverse effects on the environment from growing GM herbicide-tolerant maize compared with conventional maize
⇨ but that there would be adverse effects from growing GM herbicide-tolerant oilseed rape and beet compared with their conventional equivalents.
ACRE also recommended that:
⇨ any future commercial cultivation of the GM maize should be restricted to the conditions under which it was grown in the FSEs (i.e. the same herbicide management regime), or conditions that have been shown not to result in adverse effects
⇨ further work be conducted to investigate the implications of the impending withdrawal of atrazine (the main herbicide used with conventional maize); and that a requirement be imposed to monitor how the management of conventional maize develops in response to the phasing out of atrazine.
The results of the FSE trials of autumn-sown oilseed rape were published in March 2005, and the conclusions reached on them were the same as for the spring oilseed rape crop.
Q. *The FSE maize results are invalid because they compared GM maize with conventional maize used with the banned herbicide atrazine.*
The FSE trials were designed in 1999 and from then until 2005 atrazine was the most commonly used herbicide on conventional maize crops (starting in 2005 its use would be phased out). The GM maize in the FSEs already had an EU consent for commercial cultivation which lapsed in October 2006. ACRE considered the position in that context.
ACRE recommended that further work should be done to explore the implications of the phasing out of atrazine relative to the FSE results and, in effect, that the GM maize

consent should not be considered for renewal after 2005 without evidence that allows a comparison between the GM maize herbicide regime and whatever replaces atrazine as the main herbicide used with conventional maize.

Q. *The impact on biodiversity of GM crops should have been compared with organic crops.*
The FSE trials compared GM management with a range of intensities of conventional crop management. Although increasing, organic crops still represent only a very small area of total UK production of the crops most likely to be grown in GM varieties for the foreseeable future. In practice, for the foreseeable future, if GM crops are grown they will replace conventional varieties not organic ones so that is the key comparison.
Last modified 1 November 2006

⇨ The above information is reprinted with kind permission from the Department for Environment, Food and Rural Affairs. Visit www.defra.gov.uk for more information.
© Crown copyright

Why GM?

Information from GM Nation?

Why genetically modify food?

Sharp differences in perspective on the need for GM, and what benefits and costs it brings, lie at the heart of the GM debate.

Views for

Current GM crops provide environmental, economic and indirect health benefits. In the future they will provide direct health benefits as well.

Sharp differences in perspective on the need for GM, and what benefits and costs it brings, lie at the heart of the GM debate

---

It is important to evaluate and develop GM crops that will help support the world's population in a truly sustainable manner and to help farmers in this country and elsewhere to contribute to this goal.

GM crops can benefit the environment by reducing the needs for pesticides and fossil fuels.

Future GM crops that can be grown under environmental stresses (heat, cold, drought) will help countries (including developing countries) to improve their food security in a way that is affordable and less damaging to the environment.

Views against

GM will not 'feed the world'. The current food crisis is a problem of distribution not quantity.

The evidence so far does not show GM crops lead to reduced use of chemicals.

Anything that GM can do, other methods can also do, without bringing risks to the environment.

The debate has to be widened beyond a focus on GM alone, to look at the whole question of how we produce our food.

Who's driving GM technology?

Both sides agree that GM technology is being driven by industry and science, but disagree on their motives.

China provides the exception to this rule with most of its current GM products being developed by government.

Of the GM crops in the pipeline and coming to market the vast majority are being developed by small to medium-sized enterprises, universities, governments and other non-corporate institutions.

Views for

Science, industry and consumer demands are driving GM.

More and more farmers each year choose to grow GM crops internationally because GM allows them to grow safe, quality food crops with greater profits, whilst improving the environment in which they live and work.

People have always considered it legitimate and ethically defensible to improve natural systems for the benefit of mankind. They do not believe it is immoral for companies to make a profit, nor to gain from the investments they make in GM research and development.

Views against

A self-interested alliance of science and commerce is driving GM technology forward.

For the companies, the profit motive dominates other considerations and leads to short-term, simplistic thinking, downplaying potential threats to health and the environment.

Governments support GM because they fear their country will be left behind in the biotechnology race.

Scientists take too narrow a view of what they are doing.

Who really benefits?

Both sides agree that, today, GM companies are one of the beneficiaries, but disagree on who else benefits.

Views for

We all stand to gain from GM, and there is a real, not manufactured need for it. Healthier food, more efficiently produced, with less wastage and less use of agrochemicals and fossil fuels will be of benefit to us all.

More efficient food production means less land will be under the plough, allowing us to preserve more natural habitats.

Economic benefits accrued from current GM crops are shared between the farming community, the technology providers, seed companies, food producers and the consumer.

The indirect benefits for society at large and the environment flow

GM crop area by country

Biotech crop area by country, 2006

Country	Area of GM crops	Country	Area of GM crops
United States	54.6 milion hectares	Romania	115,000 hectares
Argentina	18 million hectares	Mexico	60,000 hectares
Brazil	11.5 million hectares	Spain	60,000 hectares
Canada	6.1 million hectares	Colombia	30,000 hectares
India	3.8 million hectares	France	5,000 hectares
China	3.5 million hectares	Iran	4,000 hectares
Paraguay	2 million hectares	Honduras	2,000 hectares
South Africa	1.4 million hectares	Czech Republic	1,290 hectares
Uruguay	400,000 hectares	Portugal	1,250 hectares
Philippines	200,000 hectares	Germany	950 hectares
Australia	200,000 hectares	Slovakia	30 hectares

Source: International Service for the Acquisition of Agri-Biotech Applications (18.1.07) (http://www.isaaa.org). Taken from information provided on the CropGen website (www.cropgen.org)

from more affordable, sustainable, environmentally friendly methods of food production.

Views against

It is chiefly the GM companies who stand to gain from GM, and perhaps some producers.

GM brings no benefits to consumers, nor indeed to most farmers. In the long term, the costs of GM, especially to the environment, will far outweigh any benefits.

The question we should really ask is not who benefits, but who *should* benefit?

Will consumers benefit?

People disagree whether ordinary consumers can really benefit from GM, now and in the future.

Views for

There are direct and indirect benefits to consumers from GM products. The current pressure to remove them from sale means that the consumer is denied choice.

Today, consumers demand safe, high quality, affordable food produced in a more environmentally friendly way. GM crops are able to supply one solution to these needs.

In the future, direct health and even greater environmental benefits will be available.

Wider social and economic benefits are realised by encouraging the growth of a very important industry.

Finally, by having a healthy farming and manufacturing sector, we can all benefit as workers and tax payers.

Views against

The question, 'will consumers benefit?' depends on what you consider a 'benefit'. We need to answer that question as a society, not as individuals.

The consumer does not benefit from GM at the moment, and will not do so in the future.

GM crops are designed to help the agriculture industry, not the consumer.

The advantages claimed for GM must be weighed against alternative ways of reaching the same goal – such as a healthier diet, less use of agricultural chemicals. Organic methods of farming offer solutions to both.

Will it make life easier?

People disagree whether GM technology will deliver on the promise of better and/or cheaper food.

Views for

People in the developed and developing world are already seeing benefits in terms of better crops which cost less to produce.

GM is also helping subsistence farmers to move out of the poverty trap.

GM crops produce higher yields using less inputs and so result in lower costs of production. Some of these savings have been shared with the consumer.

Current GM crops such as insect-protected maize have reduced damage and reduced mycotoxins levels which cause various human and animal diseases. This benefits the health of humans and animals that consume this maize.

Potential future GM crops include vitamin and nutrient-enriched foodstuffs, and crops with improved oil profiles that will help reduce coronary diseases.

However, research and development take time and these products will not be available tomorrow. Only with support for well-regulated research and development will they ever become a reality.

Views against

There is nothing that GM foods can offer that a good basic diet doesn't already offer. Manipulating the nutrient level of some foods is not a good solution – and could be dangerous.

There is no evidence that food will be cheaper. We already know how to improve our diet, even if we do not act on that knowledge.

Most of the health-enhancing crops have not come to anything yet and – even if they do – we need to balance possible advantages against obvious risks to our health and the environment.

Will GM benefit the world?

People are sharply divided on whether GM technology brings real advantages across the world, especially to developing countries.

Views for

Those generally for GM feel the key advantage that GM offers is the ability to continue to provide enough food for the world's population, which they regard as one of the fundamental challenges facing us.

They believe that if GM crops are not grown – helping to increase yield per hectare – then more and more land will have to come under the plough, destroying natural habitats.

Views against

The answer is simple – there are no advantages to the world. Claims of benefit to the developing world in particular are highly questionable.

The cause of widespread hunger is not that we are not producing enough food – India has millions of tonnes of surplus grain, for example, yet millions of people go hungry.

The promised benefits haven't arrived, and examples such as Golden Rice (rice modified to contain increased vitamin A) neither work nor address the wider problems of malnutrition in Asia.

There are alternatives that could bring equal benefits without the unknown risks associated with GM and at lower cost.

What are GM medical benefits?

People disagree about whether new GM crops that claim to improve health will work, and if the crops themselves can be safely grown.

Views for

Current GM crops are producing indirect medical benefits by providing safer, high quality affordable foods.

In the future, GM technologies will bring direct medical benefits, by way of healthier, more nutritious food.

GM is being used to develop healthier oils which can help to reduce coronary diseases. For example, by increasing the oleic acid content of soybean oil or with GM tomatoes that contain higher levels of flavanoids, powerful antioxidants which protect against cardiovascular disease.

As well as enhancing the nutritional and protective value of foods, crops can be modified to remove allergens, e.g. in peanuts.

GM also makes it possible to produce pharmaceuticals more widely available to all by reducing the costs of producing them.

Views against
GM crops offer no health advantages over a healthy balanced diet, and their potential has been over-hyped.

Problems have already arisen where residues from GM 'pharmaceutical' crops have contaminated food crops grown afterwards in the same field. And the potential for contamination means that people could be taking in inappropriate medicines without knowing it.

The debate should be widened to look at the whole issue of how we tackle our long-term health problems through diet. Promoting local fresh vegetables and fruits would bring greater benefits to UK agriculture than GM.

Will GM help feed the world?

People disagree about whether GM crops will help solve the world food crisis, or whether GM is just a 'technical fix' for a much more fundamental problem.

Food now and in the future is a global problem. The world's population is expected to rise from the present 6 billion to between 9 and 10 billion. 800 million people are already malnourished, and suffer from the diseases of malnutrition. Arable land is being lost through poor management, or built over. Fresh water supplies are in crisis.

Views for
GM offers a valuable tool with which to tackle these problems. It would be wrong to claim it is the only solution. It has the ability to transform food supplies: by reducing losses through pests and disease and through decay during storage; and by increasing nutritional value.

People disagree about whether GM crops will help solve the world food crisis, or whether GM is just a 'technical fix' for a much more fundamental problem

---

GM crops offer vital insurance against pest and insect damage. An estimated 40 per cent of developing world crops are lost to pest damage each year. GM technology cuts pest damage losses, cuts spending on pesticides and improves yield, providing a consistent, reliable insurance policy against crop failure.

Views against
GM is a 'technical fix' for a much more profound problem: people are hungry not because there is not sufficient food to feed them, but because of its unequal distribution between rich and poor. GM addresses the symptoms, not the cause.

Increasingly sophisticated methods of sustainable farming already offer a viable solution, by increasing yield.

GM crops would have a negative rather than a positive effect, if the GM companies owned patents on staple food crops.

What about non-food crops?

Agriculture does not just produce food. It also includes such things as cotton, hemp, industrial oils and fuel. Recent innovations include biofuels and plastics made from starch not petroleum. Plants that manufacture new medicinal substances are also being developed.

People tend to agree that non-food crops may be less harmful to human health, but disagree about their potential to harm the environment.

Views for
Currently, GM crops are allowing farmers to reduce the impact of farming by using less pesticides and fossil fuels. Also, new farm management practices made possible with GM crops, such as reduced cultivation techniques, can help farmers to further enhance farmland wildlife.

There is great potential in such crops to replace fossil fuels with biofuels and biodegradable starch-based plastics that break down easily in the soil and reduce greenhouse gases so helping to protect the environment even more.

The use of plants as 'factories' for pharmaceutical and industrial production also offers huge potential.

Views against
While health risks might be less (this is not proven), the risks to the environment of GM non-food crops may be as great as with food crops. They may still contaminate food crops, affect ecosystems and possibly our health (through breathing in pollen).

Environmental risks are irreversible, and are therefore especially dangerous. There is an urgent need for research into alternatives.

⇨ The above information is reprinted with kind permission from GM Nation? Visit www.gmnation.org for more information.

© GM Nation?

GM labelling

Information from the Food Standards Agency

The Agency supports consumer choice. We recognise that some people will want to choose not to buy or eat genetically modified (GM) foods, however carefully they have been assessed for safety.

Will the label tell me if food is GM?

In the EU, if a food contains or consists of genetically modified organisms (GMOs), or contains ingredients produced from GMOs, this must be indicated on the label. For GM products sold 'loose', information must be displayed immediately next to the food to indicate that it is GM.

In the EU, if a food contains or consists of genetically modified organisms (GMOs), or contains ingredients produced from GMOs, this must be indicated on the label

---

On 18 April 2004, new rules for GM labelling came into force in all EU Member States.

The GM Food and Feed Regulation (EC) No. 1829/2003 lays down rules to cover all GM food and animal feed, regardless of the presence of any GM material in the final product.

This means products such as flour, oils and glucose syrups will have to be labelled as GM if they are from a GM source.

Products produced with GM technology (cheese produced with GM enzymes, for example) will not have to be labelled.

Products such as meat, milk and eggs from animals fed on GM animal feed will also not need to be labelled. Details on the labelling rules can be found in the table below.

Any intentional use of GM ingredients at any level must be labelled. But there is no need for small amounts of GM ingredients (below 0.9% for approved GM varieties and 0.5% for unapproved GM varieties that have received a favourable assessment from an EC scientific committee) that are accidentally present in a food to be labelled.

The Food Standards Agency and the Department for Environment, Food and Rural Affairs consulted over a 12-week period on draft documents that described the scope of the new rules and draft domestic legislation which will provide penalties for enforcing the EC regulations. This consultation ended on 25 June 2004.

Following this consultation, comments received from stakeholders were analysed and any necessary changes made to the domestic legislation. The national regulations in England for food and feed came into force on 4 October 2004. Copies are available from Her Majesty's Stationery Office (HMSO). Similar regulations will apply in Scotland, Wales and Northern Ireland. A separate SI on traceability and labelling came into force on 8 October 2004.

⇨ The above information is reprinted with kind permission from the Food Standards Agency. Visit www.food.gov.uk for more information.

© Crown copyright

Guidance notes to GM labelling regulations

GM – genetically modified GMM – genetically modified micro-organism

GMO-type	Hypothetical examples	Labelling required since 18 April 2004
GM plant	Chicory	Yes
GM seed	Maize seeds	Yes
GM food	Maize, soybean, tomato	Yes
Food produced from GMOs	Maize flour, highly refined soya oil, glucose syrup from maize starch	Yes
Food from animals fed GM animal feed	Meat, milk, eggs	No
Food produced with help from a GM enzyme	Cheese, bakery products produced with the help of amylase	No
Food additive/flavouring produced from GMOs	Highly filtered lecithin extracted from GM soybeans used in chocolate	Yes
Feed additive produced from a GMO	Vitamin B2 (Riboflavin)	No
GMM used as a food ingredient	Yeast extract	Yes
Alcoholic beverages		Yes
Products containing GM enzymes where the enzyme is acting as an additive or performing a technical function		Yes
GM feed	Maize	Yes
Feed produced from a GMO	Corn gluten feed, soybean meal	Yes
Food containing GM ingredients that are sold in catering establishments		Yes*

* The FSA's legal view is that labelling is required across EU Member States under EC Regulation 1829/2003. However, there is disagreement between the Commission and the Council as to whether labelling is required.

Genetic modification and the environment

Information from the Natural Environment Research Council

Genetically modified organisms (GMOs) are a contentious issue. Many people point to the benefits GMOs could offer medicine, agriculture, and pest control. Others see GMOs as threats to the environment and to human health.

Why produce GMOs?

GM technology offers:

⇨ pest and disease-resistant crops which should lead to reductions in insecticide use

⇨ weedkiller-resistant crops that make weed control easier

⇨ rice with added vitamin A

⇨ potatoes with more protein

⇨ drought-resistant crops

⇨ plants that can produce novel products such as plastics

⇨ new treatments for genetic disorders

⇨ bacteria that can clean up soil contamination.

Set against these benefits are potential risks. Will genes from modified crops escape into wild plants, protecting them from their natural pests, or from weedkillers? Will GM crops disturb natural ecosystems, harm wildlife, or pollinate organic crops, invalidating their organic status?

In this article, we examine what we know about genetic modification of crops, and also the uncertainties that could present risks to the environment.

Selective breeding versus genetic modification

For centuries humans have changed organisms' genetic make-up by choosing individual plants and animals with particular traits, like fast growth rates or good seed production, and breeding from them. This is selective breeding, and is, in a sense, similar to evolution by natural selection. During this process

NATURAL ENVIRONMENT RESEARCH COUNCIL

thousands of genes are transferred. But selective breeding happens only within closely related species.

Genetic modification can overcome species barriers, allowing genes to be used in new ways. It also allows us to alter an organism's DNA with much greater precision – genes can be transferred or manipulated singly.

Concerns about GMOs

Could a modified organism have undesirable characteristics, for example, by becoming invasive, persistent or toxic?

Yes, it is possible. Therefore, GMOs must be carefully designed, screened and assessed for risks to health and the environment before being widely used. Genes which are known to have adverse effects are not used in GMOs. A gene introduced into corn, to make the plants more tolerant to corn-boring insects, seemed to affect lacewings which feed on the corn-borers. In laboratory trials, the lacewings showed reduced growth rates and survival. This genetic modification was abandoned. The Advisory Committee on Releases to the Environment (ACRE) advises the UK government on regulating and releasing genetically modified plants and animals.

GM terms

DNA – deoxyribonucleic acid
All living cells contain DNA – the organism's blueprint. The DNA molecule carries instructions for making all the structures and materials an organism needs to function, including the information required to instruct which genes are switched on, when, and to what extent.

Genes
Genes are segments of DNA that regulate biological activity. Genes contain the instructions for producing proteins, which make up the structure of cells and direct their activities.

GMO – genetically modified organism
A GMO is an organism that has had its DNA altered for a particular purpose. The GMO could be a virus, bacterium, plant or animal. Usually, a small section of DNA from one organism is introduced into the DNA of another with which it would not normally interbreed.

Indoor (contained) GMOs
GMOs are already used in the UK to produce medicines such as antibiotics, painkillers, vaccines, insulin and growth hormones, and some foods that rely on bacteria (such as some cheese and yoghurts). The Health and Safety Executive regulate these 'indoor' or contained GMOs. Because of stringent laboratory regulations, they are unlikely to threaten the environment.

Outdoor (released) GMOs
So far, GM crops have been grown in Britain only in tightly-regulated experiments. GM crops are not yet grown commercially in the UK but worldwide, more than eight million farmers in 16 countries grow GM crops, accounting for more than a quarter of all land under cultivation. The four main crops grown are soya beans, maize, cotton and oilseed rape.

Can transferred genes escape from GM crops into other plants?
Yes, if the GM plant can breed with wild relatives (hybridisation) and produce offspring that can themselves reproduce. Hybrids may become a problem if the plant disrupts the ecological balance in the environment, although no such effects have yet been demonstrated. Engineered genes can be genetically marked so that the frequency of gene transfers can be measured. Scientists across the globe have proposed measures that help prevent hybridisation, ranging from isolating GM crops, to using technology to ensure that hybrids either don't germinate or are infertile – terminator technology. These techniques are highly effective when implemented properly and environmental damage isn't inevitable even if GM hybrids grow.

Does it matter if genes do escape?
It depends on the nature of the trait given to them. Crops tolerant to drought or high salt concentrations could produce invasive hybrids and out-compete native plants. Plants given genes for weedkiller resistance could breed with wild relatives and produce weeds that are difficult to control amongst crops, but research so far indicates that these hybrids would not out-compete neighbours.

Can farmers grow GM crops close to conventional crops?
The distance depends on the characteristics of the particular GM crop and the level of purity required in the conventional crop. Research suggests that in some cases GM crops could be grown 100 metres from conventional crops with negligible cross contamination.

Will weedkiller-tolerant crops reduce wildlife biodiversity?
Possibly. If crop plants are resistant to more weedkillers, farmers can cultivate fields with even fewer weeds, and this will reduce biodiversity. However, this is not always true: herbicides used on conventional maize crops kill more weeds than those used with GM maize. Also, it is possible to use GM crops in farming systems that promote biodiversity.

Is there a risk of introducing antibiotic resistance via GMOs?
No. Scientists have introduced antibiotic resistance as well as the new trait into a plant during the development stage, as this is an easy way to see if the genetic modification has been successful – plants that die on exposure to the antibiotic have not been genetically modified. But antibiotic resistance has to be removed before the plants are marketed.

What are the long-term effects of GM technology on the environment?

For centuries humans have changed organisms' genetic make-up

Genetic modification of crops and micro-organisms could lead to changes such as pesticide resistance in insects – this also happens with conventional crops and with chemical pesticides.

GM technology is a further intensification of agriculture and agriculture has already changed and sometimes damaged the environment. Many of the risks people associate with GM crops are more to do with good regulation than scientific uncertainties.

Can genes inserted into one particular plant find their way into soil organisms?
Scientists have found no evidence that genes can move from plants to soil bacteria outside the laboratory.

Farm-scale crop trials
The world's largest environmental impact study of genetically modified crops ended in 2005. Over six years, scientists led by the UK's Centre for Ecology & Hydrology monitored biodiversity in 273 sites planted with GM herbicide-tolerant oilseed rape, beet and maize.

The results showed that GM beet and oilseed rape, as managed in the trials, were worse for wildlife than conventional varieties, but that GM maize was beneficial for wildlife compared to conventional maize. The results were not caused by the GM plants themselves but by the different weedkiller regimes. The team found that the beet and oilseed rape fields contained fewer of the weed seeds that provide food for wildlife, and they discovered fewer bees and butterflies used GM beet crops because there were fewer flowering weeds to provide nectar.

In the trials, conventional maize harboured the least plant and animal life, with fewer weeds, seeds and insects than GM maize.

New research suggests that farmers could modify the weedkiller regime used on GM sugar beet crops in such a way that the sugar beet can cope with more weeds than conventional sugar beet without loss of yield.

No crops have yet been approved for commercial growing in the UK.

What is the future for GM crops?
An expanding world population needs either increases in productivity by current farming systems or expansion of agriculture which could have even more damaging impacts on the environment.

The public reaction in the UK, coupled with the results from the farm-scale crop trials, means it is unlikely that GM crops will be grown in the UK in the next few years. Elsewhere it is a different story. Farmers planted 81 million hectares (or 200 million acres) of GM crops worldwide in 2004, up from 67.7 million hectares (or 167 million acres) in 2003.

Land use in Europe is changing. Scientists believe there will be an increase in crops used for biofuels, pharmaceuticals and plastics. Some of these crops may be genetically modified. More work needs to be done to see how this will change the landscape and biodiversity.

What controls are in place?
Before companies are able to market or cultivate a GMO, independent scientists must carefully screen and assess the GMO on a small scale and in an ecological context. All research activity into GMOs in the UK must be registered.

Case histories
Natural born killers, but slow
Insects have their own diseases – natural viruses known as baculoviruses. These viral infections take time to kill pests, so serious crop

damage can occur before the control measures take effect. The viruses can be made to produce a fast-acting toxin by inserting a gene from another organism. For example, a naturally occurring baculovirus which kills caterpillars has been modified with a gene from a scorpion to enable it to kill the caterpillars of the cabbage white butterfly quickly. No such GM viruses have so far received approval for commercial use.

Plastic plants

Many plants produce natural fatty acids which accumulate in their seeds as energy stores. Plants can be genetically modified to produce more of particular fatty acids. For example, researchers are modifying oilseed rape to produce chemicals used in making biodegradable plastics, and coriander is being modified to produce more petroselinic acid. This acid is used in making detergents and nylon.

Dirt busters

Bacteria and plants are often used to help clean up contaminated land and water. An example is a bacterium called *Pseudomonas* that is good at eating up waste explosives. Efforts are under way to modify certain plants and bacteria to make them more effective in dealing with particular contaminants. But at present, no modified dirt-busting bacteria are authorised for use.

⇨ The above information is reprinted with kind permission from the Natural Environment Research Council. For more information, please visit their website at www.nerc.ac.uk

© NERC

Scientists create healthier tomatoes

Scientists genetically engineer tomatoes with enhanced folate content

Leafy greens and beans aren't the only foods that pack a punch of folate, the vitamin essential for a healthy start to pregnancy.

Researchers now have used genetic engineering – manipulating an organism's genes – to make tomatoes with a full day's worth of the nutrient in a single serving. The scientists published their results in this week's online edition of the journal *PNAS*, *Proceedings of the National Academy of Sciences*.

'This could potentially be beneficial worldwide,' said Andrew Hanson, a plant biochemist at the University of Florida at Gainesville who developed the tomato along with colleague Jesse Gregory. 'Now that we've shown it works in tomatoes, we can work on applying it to cereals and crops for less developed countries where folate deficiencies are a very serious problem.'

Folate is one of the most vital nutrients for the human body's growth and development, which is why folate-rich diets are typically suggested for women planning a pregnancy or who are pregnant. Without it, cell division would not be possible because the nutrient plays an essential role in both the production of nucleotides – the building blocks of DNA – and many other essential metabolic processes.

Deficiencies of the nutrient have been linked to birth defects, slow growth rates and other developmental problems in children, as well as numerous health issues in adults, such as anaemia.

'Folate deficiency is a major nutritional deficiency, especially in the developing world,' said Parag Chitnis, programme director in the National Science Foundation's Division of Molecular and Cellular Biosciences, which funded the research. 'This research provides the proof-of-concept for the natural addition of folate to diet through enhancement of the folate content of fruits and vegetables.'

The vitamin is commonly found in leafy green vegetables like spinach, but few people eat enough produce to get the suggested amount of folate. So, in 1998, the Food and Drug Administration made it mandatory that many grain products such as rice, flour and cornmeal be enriched with a synthetic form of folate known as folic acid.

Folate deficiencies remain a problem in many underdeveloped countries, however, where adding folic acid is impractical or simply too expensive.

'There are even folate deficiency issues in Europe, where addition of folic acid to foods has not been very widely practised,' Gregory said. 'Theoretically, you could bypass this whole problem by ensuring that the folate is already present in the food.'

Will doctors be recommending a healthy dose of salsa for would-be pregnant women any time soon? Probably not, the researchers say.

'It can take years to get a genetically-engineered food plant approved by the FDA,' Hanson said. 'But before that is even a question, there are many more studies to be done – including a better look at how the overall product is affected by this alteration.'

And there is another hurdle the researchers must clear. Boosting the production of folate in tomatoes involved increasing the level of another chemical in the plant, pteridine. Little is known about this chemical, which is found in virtually all fruits and vegetables.

5 March 2007

⇨ The above information is reprinted with kind permission from the National Science Foundation. For more information, please visit the National Science Foundation's website at www.nsf.gov

© *National Science Foundation*

Biotechnology: growing in popularity

Information from the European Commission

Better informed than they were fifteen years ago, Europeans are demonstrating a greater degree of confidence and optimism in respect to the progress of biomedicine and industrial biotechnologies. Nevertheless, they remain mainly opposed to 'genetic modification, or cloning' in agriculture, and therefore, also in food. George Gaskell of the London School of Economics, Scientific Coordinator of the analysis of the sixth Eurobarometer Biotechnology survey (2005), takes us through these trends.

Who knows nothing about stem-cell research, progress in prenatal diagnosis, the problems of co-existence of GM and traditional agriculture, or the trivialisation of genetic fingerprinting? Anything that comes into contact with rapid advances in biological sciences is now the subject of an intense flow of information and debates in the media that are registering higher and higher in public perception.

The distrusting nineties

'Europeans are displaying increasing interest in and attention to these questions, which they are encountering more and more in their daily lives, particularly in relation to looking after their own health,' comments George Gaskell of the London School of Economics, Scientific Coordinator of the Eurobarometer Biotechnology survey. In contrast to the elevated levels of optimism that accompanied advances in sectors such as IT or solar technologies, those relating to bio-innovation saw a relative level of distrust among public opinion during the 1990s. The announcement of the development of genetically modified strains, i.e. food, and the birth of Dolly the sheep definitely troubled many minds that had already been shaken by the BSE crisis.

Despite the popularisation in the media of anti-GM movements, the under-25s is the group that is mostly prepared to consume GM produce when it comes on the market

---

'Today, biosciences are regaining the mostly optimistic perception they previously enjoyed, particularly in those areas we refer to as the red biotechnologies – linked to medicine and healthcare – and also the white biotechnologies – industrial-based uses such as bio-fuels, bio-plastics and bio-pharmaceuticals. The exception to this trend is seen in the distrust, if not out-and-out rejection, of green biotechnologies, which relate to genetic manipulation in agriculture – and therefore, by extension, food – and/or the natural environment.'(1)

Internal vs external

This resistance is unique, and indubitably merits some thought. Why is public opinion more accepting of innovation in genetic engineering in relation to taking care of the inside of the human body than when we try to use these techniques to change that which we grow outside our bodies to feed ourselves?

'I believe in a realist interpretation of this behaviour. The majority of people recognise more and more that we should take – under the guarantee of scientific expertise and ethical

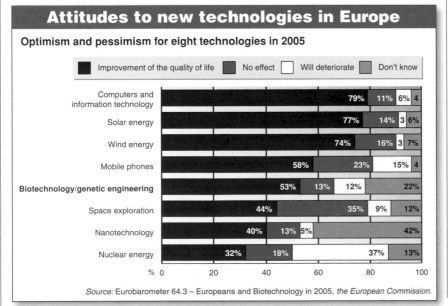

Attitudes to new technologies in Europe

Optimism and pessimism for eight technologies in 2005

Legend: Improvement of the quality of life | No effect | Will deteriorate | Don't know

Technology	Improvement of the quality of life	No effect	Will deteriorate	Don't know
Computers and information technology	79%	11%	6%	4
Solar energy	77%	14%	3	6%
Wind energy	74%	16%	3	7%
Mobile phones	58%	23%	15%	4
Biotechnology/genetic engineering	53%	13%	12%	22%
Space exploration	44%	35%	9%	12%
Nanotechnology	40%	13%	5%	42%
Nuclear energy	32%	18%	37%	13%

Source: Eurobarometer 64.3 – Europeans and Biotechnology in 2005, *the European Commission.*

supervision – the calculated risks that are inherent to bio-innovation, where they perceive a promising potential use. Nevertheless, they also want us to prove the tangible advantages of the biological prowess and promise that scientists and industry are striving to promote. These people are optimists in the best way – pragmatic and prudent. In the case of GM organisms and their impact on food production and nature, these people have doubts concerning the interests of even the innovations whose praises are being sung. Apart from the argument of fighting famine, why should we have to change, right now, the way in which nature supplies and always has supplied our food?'

Beyond GM organisms, another hot topic explored by the survey is the perception of opinion on stem-cell research. What are the ethical problems raised by this research, especially in relation to any specific religious beliefs that may or may not be held by those interviewed? Even among those who attend a place of worship, realistic expectations of the positive findings of this research in the medical field dominate, as long as there is control of the consequences of the research on morals and society. 'The question is whether the promise attributed to advances in stem-cell research, largely disseminated through the media, is realistic, or whether it borders on hyperbole. If the latter is the case, the optimistic viewpoint risks disappearing.'

Young people: what generational change?

The Eurobarometer also allows analysis of attitudes towards live human research on the basis of the age of respondents. 'There are frequently statements

---

Bio-innovation saw a relative level of distrust among public opinion during the 1990s. The announcement of the development of genetically modified strains and the birth of Dolly the sheep definitely troubled many minds

---

about the worrying lack of interest in science, even an opposition to it, among young people. That is not what comes out of this survey, however. They have a positive opinion that is usually equivalent to, and occasionally greater than, that of the next class of adults (from 25 to 45 years of age), which extends to practically all subjects.' Therefore, despite the popularisation in the media of anti-GM movements, the under-25s is the group that is mostly prepared to consume GM produce when it comes on the market. Is this the manifestation of a generational change by a class of young people brought up since infancy in the world of fast food, and no longer encumbered by the nutritional neuroses of their elders? As to the question of whether they worry about the long-term effects of their diet on their health, three out of five young people said they were not or only a little concerned. This indifference seems to represent the expression of a fairly typical adolescent attitude, rather than a real change in mentality. 'This result is particularly gloomy when you consider the problems presented by the emerging preoccupation with childhood obesity...'
(1) All quotes are from George Gaskell.
December 2006

⇨ The above information is reprinted with kind permission from the European Commission. Visit http://ec.europa.eu for more information.
© *European Commission*

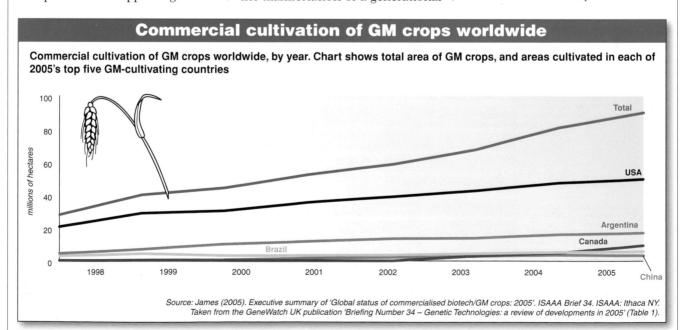

Commercial cultivation of GM crops worldwide

Commercial cultivation of GM crops worldwide, by year. Chart shows total area of GM crops, and areas cultivated in each of 2005's top five GM-cultivating countries

Source: James (2005). Executive summary of 'Global status of commercialised biotech/GM crops: 2005'. ISAAA Brief 34. ISAAA: Ithaca NY. Taken from the GeneWatch UK publication 'Briefing Number 34 – Genetic Technologies: a review of developments in 2005' (Table 1).

Statistics taken from the EC document *Europeans and Biotechnology in 2005: Patterns and Trends (Eurobarometer 64.3)*, published May 2006

Support for four technologies

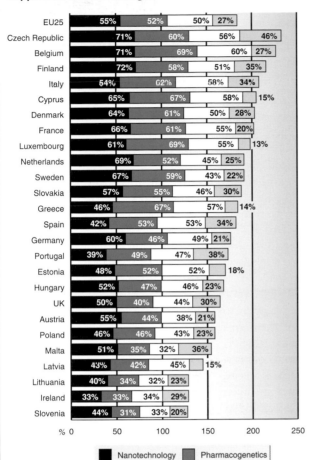

Source: Eurobarometer 64.3, Figure 5

Reasons for buying or not buying GM foods

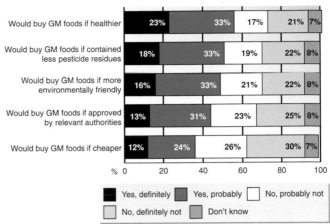

Source: Eurobarometer 64.3, Figure 7

Trends in optimism for biotechnology in the UK and Ireland

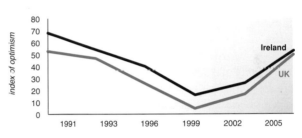

Source: Eurobarometer 64.3, Table 1

Age and intentions to purchase GM food

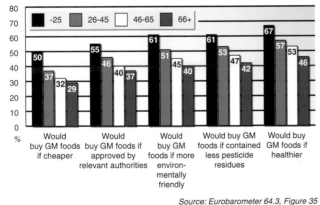

Source: Eurobarometer 64.3, Figure 35

Participation in issues concerning biotechnology

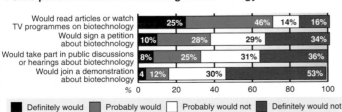

Source: Eurobarometer 64.3, Figure 26

Participation in issues concerning biotechnology

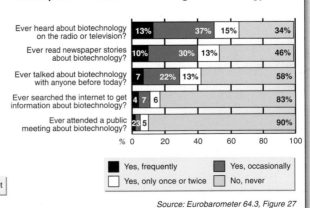

Source: Eurobarometer 64.3, Figure 27

UN upholds moratorium on terminator seed technology

Worldwide movement of farmers, Indigenous peoples and civil society organisations calls for ban

It's official. Governments at the United Nations Convention on Biological Diversity (CBD) have unanimously upheld the international de facto moratorium on Terminator technology – plants that are genetically engineered to produce sterile seeds at harvest. The 8th meeting of the CBD ended today in Curitiba, Brazil.

'The CBD has soundly rejected the efforts of Canada, Australia and New Zealand – supported by the US government and the biotechnology industry – to undermine the moratorium on suicide seeds,' said Maria Jose Guazzelli of Centro Ecológico, a Brazil-based agro-ecological organisation.

'By consensus decision, all governments have reaffirmed the moratorium on a genetic engineering technology that threatens the lives and livelihoods of 1.4 billion people who depend on farmer-saved seed,' said Pat Mooney, Executive Director of ETC Group.

Over the past two weeks, the call for a ban on sterile-seed technology took centre stage at the UN meeting in Brazil. Thousands of peasant farmers, including those from Brazil's Landless Workers Movement (Movimento Sem Terra) protested daily outside the UN meeting to demand a ban, and the women of the international Via Campesina movement of peasant farmers staged a powerful silent protest inside the meeting on 23 March.

'Terminator seeds are genocide seeds,' said Francisca Rodríguez from Via Campesina. 'We have pride in being one more step forward in our struggle but we will not stop until Terminator is banned from the face of the earth.'

The CBD's moratorium on Terminator, adopted six years ago, was under attack by three governments – Australia, Canada and New Zealand – that insisted on a 'case-by-case risk assessment' of the technology. A broad coalition of farmers, social movements, Indigenous peoples and civil society organisations pressed governments meeting in Brazil to reject the controversial text because it threatened to open the door to national-level field testing of Terminator, without regard for its devastating social impacts.

On 23 March, Malaysia, speaking on behalf of the G77 and China (together a group of 130 developing nations), said that the reference to case-by-case risk assessment was 'clearly unacceptable' because it would potentially allow field tests. Today the CBD reaffirmed the moratorium on Terminator and even strengthened it by making it clear that any future research would only be conducted within the bounds of the moratorium – meaning no field trials.

Leading up to the UN meeting, civil society groups and social movements across the globe intensified their campaigns against Terminator – sending a strong message to governments meeting in Brazil. Actions include:

⇨ In India, farmers collected over a half-million signatures calling on the Prime Minister to remain strong in defending the national ban on Terminator and upholding the international moratorium;

⇨ On 16 March, the European Parliament passed a resolution calling on European governments to uphold the CBD moratorium and reject text on 'case by case';

⇨ On March 23, following extensive consultations, Indigenous community leaders in Peru called on multinational company Syngenta to abandon its Terminator-like patent on potatoes;

⇨ In Madrid on March 23, anti-Terminator protesters planted local varieties of organic vegetable seeds outside Monsanto's offices;

⇨ Last week groups targeted those countries supporting Terminator and, in addition to domestic letter-writing campaigns, protests were held at the New Zealand embassies in London and New Delhi, and a protest was held at the Canadian embassy in Berlin.

'The international moratorium on Terminator has been upheld – but the battle isn't over yet. Terminator will be commercialised unless national governments take action to ban it – as Brazil and India have done,' said Lucy Sharratt of the international Ban Terminator Campaign.

5,000 peasant farmers protested to-day outside the UN conference to send government delegates home with their message to protect farmers' rights. *31 March 2006*

⇨ Information from Ban Terminator. Visit www.banterminator.org for more information.

© Ban Terminator

GURTs

'Terminator seeds' are one example of plants bred using Genetic Use Restriction Technologies (GURTs). Plants bred with this type of GURT – also known as a v-GURT – produce sterile seeds so that a farmer would not be able to use the seed from this crop for future planting. These techniques can be applied by genetic modification of a plant and by more traditional methods of plant breeding.

A second type of GURT – known as a t-GURT – modifies a crop in such a way that a specific enhanced trait in the crop is not expressed until application of a specific chemical or activator.

Source: Defra. Crown copyright.

GM contamination

First contamination report reveals worldwide illegal spread of genetically engineered crops – Greenpeace and GeneWatch UK call for urgent adoption of international biosafety standards

The first report into the extent to which genetically engineered (GE) organisms have leaked into the environment – released today – reveals a disturbing picture of widespread contamination, illegal planting and negative agricultural side effects.

The report is a summary of incidents uncovered by the online Contamination Register set up by Greenpeace and GeneWatch UK. It reveals a catalogue of highly disturbing incidents right across the world, including:

⇨ Pork meat from genetically engineered pigs being sold to consumers

⇨ Ordinary crops being contaminated with GE crops containing pharmaceuticals

⇨ Growing and international distribution of illegal antibiotic-resistant maize seeds

⇨ Planting of outlawed GE crops which have been smuggled into countries

⇨ Mixing of unapproved GE crops in food, including shipments of food aid

⇨ Inadvertent mixing of different GE strains even in high-profile scientific field trials.

The report reveals 113 such cases worldwide, involving 39 countries – twice as many countries as are officially allowed to grow GE crops since they were first commercialised in 1996. Worryingly, the frequency of these cases is increasing, with 11 countries affected in 2005 alone. Contamination has even been found in countries conducting supposedly carefully controlled high-profile farm-scale evaluations, such as the UK.

'This may well only be the tip of the iceberg, as there is no official global or national contamination register so far,' said Dr Sue Mayer of GeneWatch UK, who leads the team of investigators. Most incidents of contamination are actually kept as confidential business information by companies as well as public authorities.

The report reveals a disturbing picture of widespread contamination, illegal planting and negative agricultural side effects

---

Greenpeace is calling for a mandatory international register of all such events to be set up, along with the adoption of minimum standards of identification and labelling of all international shipments of GE crops. Without such biosafety standards, the global community will have no chance of tracing and recalling dangerous GMOs, should this become necessary, said Benedikt Haerlin of Greenpeace International's Biosafety Protocol delegation.

The publication of the report comes only days before the latest meeting of the 132 countries who have signed the Biosafety Protocol, which is to establish standards of safety and information on GE crops in global food and feed trade. At their last meeting an imminent agreement was blocked by only two member states, Brazil and New Zealand. They were backed by the major GE-exporting countries USA, Argentina and Canada, who are not members of the Protocol and want to restrict required identification to a meaningless note that a shipment 'may contain' GE.

'All of these countries have national legislation to protect themselves from illegal GE imports. Still they want to deny the same rights and level of information to less developed countries, with no national Biosafety-laws and means to enforce them,' concluded Haerlin. 'Do they really want such unethical double standards and to create dumping grounds for unidentified and illegal GE imports? We hope that Brazil, who will be hosting this meeting, will not betray the developing countries and cater to large agro-businesses at the expense of the environment.'
8 March 2006

⇨ The above information is reprinted with kind permission from GeneWatch UK. Visit www.genewatch.org for more information.

GM drug crops

Warning over GM drug crops – as US prepares to allow GM rice with human DNA

Friends of the Earth is calling for the production of drugs in food crops grown outside to be banned after the US Department of Agriculture (USDA) gave preliminary approval to the commercial production of GM pharmaceutical rice containing human genes [1]. The environmental campaign group warned of the potentially devastating consequences if pharmaceutical crops end up on consumers' plates.

The warning comes as US authorities have confirmed that a third GM rice contamination incident in less than a year has hit the United States. In the latest incident a type of non-GM long grain rice (known as Clearfield CL131, produced by BASF) was found to contain unknown GM contamination. The USDA has stepped in to stop rice farmers planting the variety because of the likelihood that the GM trait is unapproved.

Last week, US authorities confirmed that Clearfield CL131 had also been contaminated by GM LL62 rice – produced by biotech company Bayer CropScience. Because this rice is legal in the US, farmers had decided to plant the variety this spring because of a shortage of seed. This follows the initial contamination incident with Bayer's LL601 rice which affected long grain rice exported around the world, including the UK.

Friends of the Earth's GM campaigner, Clare Oxborrow, said:

'This latest GM contamination incident should set alarm bells ringing about the dangers of allowing GM pharmaceutical crops to be grown. Using food crops and fields as glorified drug factories is deeply worrying. The biotech industry has repeatedly failed to prevent experimental GM rice contaminating food crops. If pharmaceutical crops end up on consumers' plates, the consequences for our health could be devastating.'

'The UK Government must urge the US to ban the production of drugs in food crops grown outside. It must also introduce tough measures to prevent illegal GM crops contaminating our food and ensure that biotech companies are liable for any damage their products cause.'

Note

1. The USDA has given preliminary approval for the first GM pharmaceutical rice containing human genes to be grown commercially. The rice, produced by a company called Ventria, has been engineered to produce human proteins to be extracted to produce anti-diarrhoea medicine.
6 March 2007

⇨ The above information is reprinted with kind permission from Friends of the Earth. Visit their website at www.foe.co.uk for more information.

© *Friends of the Earth*

Modified hens lay eggs to help beat cancer

By Roger Highfield, Science Editor

A flock of designer hens, genetically modified with human genes to lay eggs capable of producing drugs that fight cancer and other life-threatening diseases, has been created by British scientists.

Researchers at the Roslin Institute near Edinburgh, which created pioneering GM animal 'drug factories' such as Tracy the sheep as well as Dolly the clone, have bred a 500-strong flock of ISA Browns. These are a prolific egg-laying French cross between Rhode Island Red and Rhode Island White chickens.

Because they make proteins used as drugs in the whites of their eggs, they offer the prospect of mass-producing, at a fraction of the price, drugs that cost thousands of pounds a year per patient. This marks an important advance in the use of farm animals for the production of pharmaceuticals.

Existing methods for producing protein drugs, such as monoclonal antibodies used to treat cancer and arthritis, are expensive and time-consuming.

Using GM – 'transgenic' – farm animals for the mass production of such drugs is potentially cheaper, faster, and more efficient than standard methods, but researchers so far have been unable to make 'pharming' workable.

The GM chickens are reported today in the *Proceedings of the National Academy of Sciences* by Dr Helen Sang and colleagues in the Roslin Institute and the companies Oxford BioMedica, which specialises in gene therapy, and Viragen, which is commercialising the technology.

They describe how they have produced transgenic hens by using a particular virus – an equine infectious anaemia lentivirus – to insert the genes for desired pharmaceutical proteins into the hen's gene for ovalbumin. This is a protein that makes up 54 per cent of egg whites, around 2.2 grams for each egg – a massive amount by the standards of biotechnology.

They inserted the human genes into chicken embryonic stem cells, then blended those cells with those of a normal chicken embryo to create a chimera, a blend of GM and normal cells.

Crucially, the cells in the oviduct (which lays eggs) consisted of GM cells and so passed on the implanted gene so the egg could make the drug protein.

The working proteins in these hens included miR24, a monoclonal antibody with potential for treating malignant melanoma, and human interferon b-1a, an antiviral drug.

Just as important, the genes were passed on to the next generation.

Although there have been attempts to make protein drugs in the milk of sheep, goats, cattle and rabbits, the team believes that the conversion of chickens into 'bioreactors' offers many advantages. They produce more quickly and are much cheaper to look after.

'This is potentially a very powerful new way to produce specialised drugs,' said Dr Karen Jervis of Viragen Scotland, which worked with the Roslin team.

'We have bred five generations of chickens so far and they all keep producing high concentrations of pharmaceuticals.'

Andrew Wood, of Oxford BioMedica, whose researchers collaborated on the project, said: 'This could lead to treatments for Parkinson's disease, diabetes and a range of cancers.'

15 January 2007

GM tobacco could save lives

How a tobacco farm in Kent could provide a life-saving drug for millions

⇨ *Genetic tweak allows HIV drug to be harvested*
⇨ *Environmentalists fear cross-contamination*

In the perfectly controlled atmosphere of a brick-proof, hermetically sealed greenhouse deep in the Kent countryside, a fresh crop of tobacco plants is beginning to flourish.

There is nothing unusual about the plants' appearance, but they are nonetheless extraordinary. A genetic tweak ensures that every cell of every plant churns out tiny quantities of an experimental drug. When harvested, they could bring cheap medicine to millions.

Scientists say the £8m project could provide a powerful weapon against Africa's HIV pandemic.

The process is called pharming, and to many it is both the future of GM crops, and the future of the drugs industry. If the tobacco plants in Kent are a success, each one will provide 20 doses of an anti-HIV drug – enough to protect a woman from infection for up to three months.

Pharming is a marriage of high and low technology that capitalises on the advantages of both. Instead of needing a $500m drug manufacturing facility that takes five years to pass

---

By Ian Sample, Science Correspondent

---

regulatory approval, pharming uses simple crop-growing practices that have been honed over centuries.

The process is called pharming, and to many it is both the future of GM crops, and the future of the drugs industry

---

Like other GM technologies, pharming is not without its risks. Pressure groups such as Friends of the Earth fear that if food crops such as maize or tomatoes are adopted to grow drugs in some regions, there is a risk of their contaminating maize or tomato crops elsewhere that are intended for consumption. Clare Oxborrow, FoE's GM campaigner, said: 'We wouldn't want to see this

done in food crops and certainly not in field trials.'

Professor Julian Ma, who leads the tobacco plant project at the Centre for Infection at St George's hospital in south London, acknowledges that the plants, and more importantly their pollen, have to be well contained. It is why the plants are being grown in £35,000 high-security Unigro greenhouses which normally house experiments on plant viruses. Designed to withstand a lobbed brick, the greenhouses are twin-skinned plastic. Rupture either skin and the entire greenhouse is immediately flooded with formaldehyde, keeping everything inside.

At his labs at St George's, Prof Ma and his PhD student Amy Sexton have been producing the genetically modified tobacco plants and perfecting techniques to boost the amount of drug each plant makes. They take a common tobacco plant, *Nicotiana tabacum*, and punch small holes from the leaves. The circles of leaf are placed in a petri dish and then squirted with a liquid containing a genetically modified bacterium.

When the bacterium infects a plant leaf, it inserts some of its own genes into the plant's DNA, in effect hijacking its cellular machinery, and fooling the plant to produce new proteins.

In the wild these proteins cause tumours that kill the plant. But in the laboratory, the bacterium is made safe and doctored with different genes that fool the plant into making cyanovirin-N.

The researchers believe that cyanovirin-N could become a powerful new weapon in the fight against HIV, as it prevents the virus from infecting human cells. They are keen to make a microbicide cream for women in Africa and other developing countries where many have little or no control over their partner's use of a condom. The evidence so far is that a microbicide cream could dramatically cut down the spread of HIV through sexual activity. Experiments with rhesus macaques, which have similar reproductive physiology, have shown the microbicide protected 15 out of 18 monkeys from infection with a variant of the HIV virus, while all of eight control animals were infected.

To produce enough cyanovirin to make any significant impact on the HIV pandemic will take a lot of plants. The team calculates that 5,000kg of cyanovirin would be needed for 10 million women to have two doses a week – a scale of production that is far beyond the capabilities of conventional drug manufacturing. Each plant grows to a final weight of around 1kg.

Already the team is working on ways to maximise the amount of drug it can extract from a plant. Instead of growing the plants in soil, Prof Ma is experimenting with hydroponics, where the plants are grown in a nutrient-rich liquid. 'The beauty of this is that the roots of the plant can be made to secrete the cyanovirin-N into the water they are grown in. That's a much simpler and cheaper way to extract the drug than having to grind the plants up,' he says. 'You can think of it as molecular milking.'

If the plants continue to grow well in Kent – at the home of the East Malling Research facility – Prof Ma hopes to have enough drug to conduct human clinical trials of the microbicide within three years.

Tight controls are needed to ensure that GM crops do not contaminate natural plants

'After the GM food debate, everyone was wondering, is this technology going to fly? We have here a potentially important intervention against HIV, but one that needs enormous production capacity if it's going to make an impact globally on health. GM plants could provide the solution,' says Prof Ma.

FAQ: Pharming HIV treatments
How many people have HIV?
Globally, 40m people are believed to be infected with HIV, 25m of whom live in sub-Saharan Africa. More than 40,000 people in the UK are receiving treatment for HIV. However, it is estimated a further 20,000 are infected but do not know it.

What treatments are there?
In developed countries, expensive cocktails of drugs are used to stop HIV becoming Aids. The treatment is scarce in countries most in need of it. Condoms are the most effective barriers to sexually transmitted HIV, but in many countries, women may not be able to insist on their use.

So far, none of the 90 or so experimental vaccines against HIV have proved successful, but there are high hopes for microbicide creams, which women can apply before sex. Trials of anti-HIV creams are continuing in South Africa and Uganda.

How does cyanovirin-N work?
To infect a human immune cell, the HIV virus has to latch on to the cell in a specific way. A protein on the HIV virus surface locks on to what is called the CD4 receptor on the immune cell, and from there, the virus can infect the cell. Cyanovirin works by latching on to the HIV virus, making it unable to stick to human cells.

What are the risks of pharming?
Tight controls are needed to ensure that GM crops do not contaminate natural plants. Growing in air-tight greenhouses prevents pollen escaping, but an alternative is to grow GM crops that have no relatives they can pollinate. Scientists are also working on infertile GM crops that do not flower.
4 July 2006

What are the ethics?

Information from GM Nation?

Is GM ethical?

There is no agreement over the ethical issues surrounding GM.

Views for

This issue has been carefully considered by independent ethical bodies and individuals like the Nuffield Bioethics Council, the Church of Scotland Society, Religion and Technology project, the Church of England Commissioners and the Pope. None has found any reason to consider GM technology as unacceptable in principle.

There is no agreement over the ethical issues surrounding GM

All technologies if handled properly benefit humankind. It is unethical to deny the exploration of the potential of such beneficial technologies.

Views against

Humans have always 'tampered' with nature – but GM represents a fundamental change in the way we deal with nature.

Some people believe that GM is unethical. Food containing GM material must be labelled so that these people can avoid GM if they wish.

What does the GM future hold?

At this stage of development there is no consensus on what GM will mean for future generations.

Views for

New technologies including GM offer exciting opportunities for a more sustainable future, hope of alleviating existing malnutrition and poverty around the globe, and a way of coping with the projected major increase of the world's population without mass starvation or the destruction of much of the remaining wilderness areas.

We have an obligation to evaluate all avenues that may achieve these goals.

Views against

The unknowns surrounding GM and the risks that have not yet been fully understood mean that we are leaving a deeply worrying legacy for the future. We could be posing major environmental, socio-economic and even health problems for future generations.

There is no strong evidence that GM offers a more sustainable future. In fact, it threatens to exacerbate the current problems facing the world's population.

What about the third world?

There is absolutely no agreement on this question from either side.

Views for

African countries should be allowed to decide for themselves without being dictated to by largely overfed societies.

GM crops provide a valuable opportunity for increasing food supply to malnourished people. Those in the countries concerned should be fully involved in the evaluation process.

Two-thirds of the farmers currently growing GM crops live in resource-poor countries, including those in Africa.

Using GM crops for food aid cannot be described as 'dumping' when up to 300 million people within the donating country (the US) are consuming GM crops as a normal component of their daily diet without any associated health issues.

Views against

Africa is being used as a dumping ground for GM foods. Food aid from the US is often linked to GM produce.

Long-term food security could be threatened by the use of GM crops in the West and the people who benefit are not the poor but the biotechnology companies and the seed distributors.

Is patenting genes democratic?

Views are divided on whether patenting genes is democratic.

Views for

Patenting is an essential part of the process of investing in the research and development that leads to the discovery of something new.

The patent system was introduced to enable people to share information while rewarding investment in research. Patents last about 20 years, which is a short time in the context of a plant breeding timetable. After the patent ends, everyone has access to the intellectual property.

There are well-documented examples (like Golden Rice) where companies have waived their rights to remuneration from these patents.

Views against

Gene patenting is the kind of issue that should be opened up for public debate and scrutiny. There is a real question as to whether anyone should have the right to 'own' genes.

Patenting allows industry to take control of and exploit organisms and genetic material, treating them as exclusive private property that can be sold to or withheld from farmers, breeders, scientists and doctors. For example, technology fees on seeds deprive farmers of their generations-old right to replant and exchange their seeds.

⇨ The above information is reprinted with kind permission from GM Nation? Visit www.gmnation.org for more information.

© GM Nation?

What's the problem?

Information from Greenpeace

Food is central to life. How we grow it affects the land, water and wildlife around us, as well as farm animals, our health and rural communities. It's hard to resist cheap food or 'buy one get one free' offers but the environmental cost of the culture of intensive farming, over-production and over-consumption is enormous.

Genetic engineering enables scientists to create plants, animals and micro-organisms by manipulating genes in a way that does not occur naturally, often by taking DNA from one species and inserting it into another, completely unrelated one. Jellyfish genes have been inserted into pigs, firefly genes have been bred into tobacco plants, and bacterial genes are present in crops such as soya, maize and cotton.

These genetically modified organisms (GMOs) can spread through nature and interbreed with naturally-occurring organisms, replicating themselves and spreading through the environment in an unpredictable and uncontrollable way. Their release into the environment is a form of genetic pollution and a major threat because, once they're out there in the wild, they cannot be recalled.

But because of the commercial interests of wealthy governments and biotech companies such as Monsanto and Bayer Cropscience, the public is being denied the right to know about genetically modified (GM) ingredients in the food chain and risks losing the right to avoid them.

Genetic engineering and GM foods are being sold as a way to provide crops that are disease and drought resistant as well as being able to provide more food for the world's poor. However, after decades of research there are no GM food crops that live up to all this hype. The only notable effects of GM technologies have been an increase in herbicide use and a wealth of contamination scandals, either in shipments of non-GM foods or in cross-contamination of the crops themselves.

On top of all this, the multinational biotechnology companies own all patent rights to the crop varieties they develop, increasing their stranglehold on global agriculture and allowing them to generate vast profits.

> **Genetic engineering enables scientists to create plants, animals and micro-organisms by manipulating genes in a way that does not occur naturally**

While scientific progress on molecular biology has enormous potential to increase our understanding of nature and provide new medical tools, it should not be used as a justification to turn the environment into a giant genetic experiment driven by selfish commercial interests. The biodiversity and environmental integrity of the world's food supply is too important to be put at risk.

A piece of international regulation called the Biosafety Protocol aims to regulate the use and movement of GMOs, but again biotech companies and governments sympathetic to their interests are attempting to disable it, making the familiar argument that environmental protection is a barrier to international trade.

We believe that GMOs should not be released into the environment.

Scientific understanding of their impact on the environment and human health is not adequate to ensure their safety. We also oppose all patents on plants, animals and humans, as well as patents on their genes. Life is not an industrial commodity and when we force life forms and our world's food supply to conform to human economic models rather than their natural ones, we do so at our peril.

Instead, we advocate a move away from industrial-scale agriculture towards locally-focused and sustainable models. Feeding the world without exhausting the planet's natural resources is achievable, but the food security of local communities needs to be put ahead of commercial interests.

⇨ The above information is reprinted with kind permission from Greenpeace. Visit www.greenpeace.org.uk for more information.

© Greenpeace

GM material in animal feed

Information from the Food Standards Agency

Before a GMO can either be grown or marketed in the EU, it must be granted a marketing consent under EC legislation – previously EC Directive 2001/18 on the deliberate release into the environment of GMOs, and now EC Regulation 1829/2003 laying down the authorisation procedures for GM food and feed.

This procedure applies to both living GMOs such as cereal grains, and to animal feed ingredients that are obtained by processing GM crops.

Materials from GM crops are used in animal feed in the UK, and are subject to a safety assessment as part of their authorisation. On the basis of these assessments, there is no reason to suppose that GM feed presents any more risk to farmed livestock than conventional feed. GM feed, which is very unlikely to contain viable GMOs, is digested by animals in the same way as conventional feed.

There have been some concerns that functional transgenes from GM-derived feed material might be incorporated into livestock products for human consumption (milk, meat and eggs), but research to date has failed to demonstrate any discernible occurrence of this. Food from animals fed on GM crops is therefore considered to be as safe as food from animals fed on non-GM crops.

Eight plant lines with potential use in animal feed have been licensed for commercialisation in the EU. These comprise two herbicide-tolerant and insect-resistant maize varieties (both from Syngenta), two herbicide-tolerant maizes (from Bayer and Monsanto), two insect-resistant maizes (Monsanto), a herbicide-tolerant soya bean (also Monsanto), and a herbicide-tolerant oilseed rape (Bayer). Only three of these varieties – the Bayer and one each of the Monsanto and Syngenta maizes – have been licensed for cultivation in the EU; the other five have been approved for import

and processing only. However, GM crops are commercially grown only in five EU Member States: Spain, Portugal, France, Germany and the Czech Republic.

A larger number of GM crops, including varieties of maize, soya, oilseed rape and cotton which have not received marketing consents in the EU, have been approved for growing outside the EU – particularly North and South America, South Africa, China, India and other parts of the Far East. As commodity-exporting countries have adopted GM crop technology, supplies of feed materials to the UK have increasingly contained GM products. Indications are that the trend in adoption of the technology will continue and that the proportion of GM soya beans and maize used in feed production will be maintained or increased.

In 2005, the UK imported approximately 1.8 million tonnes of soya beans and soya bean meal from the USA, Canada, Brazil and Argentina. The UK also imported approximately half that volume of maize gluten feed from the USA, plus smaller quantities of rapeseed and cotton meal from other parts of the world.

Maize and oilseed products play an important role in animal feeding. These are major sources of feed energy and protein, which would be difficult to replace. Imported soya and maize by-products (notably soya bean meal and maize gluten feed) account for approximately 20% of raw materials used by UK feed manufacturers and farmers.

Before 18 April 2004, GM material imported for feed use was not required to be labelled.

Since then, however, labelling has been required for imported feed materials which contain GM or GM-derived material. Labelling is not required for consignments containing adventitious or technically unavoidable traces of GM, up to a threshold of 0.9% for GM varieties approved in the EU and 0.5% for varieties which have received a favourable scientific assessment but have yet to be authorised in the EU.

Data on the quantities of imported feed materials which may be GM or GM-derived are not collected. However, the US Department of Agriculture estimates that GM now accounts for 52% of the US maize crop, 61% of the cotton crop and 87% of the soya bean crop. In Argentina, over 98% of soya plantings and 55% of maize plantings are GM. In Brazil, GM technology was not legally commercialised until 2005 although GM soya, estimated to account for 30% of Brazilian soya production, was being planted in the main soya-growing areas. 66% of the Chinese cotton crop is now GM.

The global area of GM crops for 2005 was 90 million hectares in 21 countries, up from 81 million hectares in seventeen countries in 2004 and 67.7 million hectares in eighteen countries in 2003. This is the tenth consecutive year of increase in the area devoted to GM crops, with much of the increase being in less developed countries. GM crops now occupy 5% of the world's cultivable arable land, an area equivalent to three times the size of the UK. Soya (almost all herbicide tolerant) and maize (2/3 insect resistant, 1/3 herbicide tolerant) account for over 80% of this.
15 May 2006

⇨ The above information is reprinted with kind permission from the Food Standards Agency. Visit www.food. gov.uk for more information.
© Crown copyright

Supermarkets supporting GM through the back door

Information from Friends of the Earth

A new survey by Friends of the Earth, reveals that most animal products sold in supermarkets, including milk, cheese and meat, come from animals fed on GM feeds. But consumers are not aware of what they are buying, with five out of 10 supermarkets failing to tell customers when food comes from animals fed on genetically modified feed.

Most animal products sold in supermarkets, including milk, cheese and meat, come from animals fed on GM feeds

---

The results come as a new GfK NOP Omnibus poll for Friends of the Earth and GM Freeze found that 87 per cent of the public think that foods from animals fed on a GM diet should be labelled. Animal products produced using GM feed are currently exempt from labelling requirements.

Friends of the Earth surveyed the 10 main supermarkets' policies on GM animal feed. The results showed that supermarkets providing the fewest non-GM-fed options were Budgens, which only sources non-GM-fed poultry and pork, Tesco, Asda, Morrisons, Iceland and Somerfield, which only source non-GM eggs, poultry and farmed fish (Morrisons additionally provides non-GM-fed New Zealand lamb).

Marks & Spencer had the widest selection of non-GM-fed products including fresh milk, meat, poultry, eggs and fish. Sainsbury's, the Co-op and Waitrose source a number of non-GM-fed products but could do more.

Friends of the Earth's GM Campaigner, Clare Oxborrow, said:

'Most people think that supermarkets are GM free, but thousands of tonnes of GM soya and maize are coming into the UK to feed the animals producing the meat, milk and other dairy products found on their shelves. Consumers clearly want to avoid food with GM ingredients but the information just is not made available.

'The intensive production of GM crops for animal feed is associated with damaging environmental and social impacts around the world. The failure of food companies to demand non-GM feed is making these problems worse – and could also result in a shortage of GM-free ingredients for food. Supermarkets must do more to phase out the use of GM animal feed, provide accurate information to help their customers choose non-GM options and contribute to finding sustainable alternatives to imported, intensively-farmed animal feeds.'

Photo: Marja Flick-Buijs

Friends of the Earth also posed as a customer and contacted the 10 main supermarkets to ask them to specify which of their products come from animals not fed a GM diet. Half of them, including Tesco and Asda, failed to answer the basic question.

The majority of GM crops grown are used for animal feed (primarily soya and maize). Environmental campaigners are concerned by the trend as growing GM crops is leading to further intensification of farming practices, environmental damage and negative social impacts. The failure to take action to replace GM in animal feed will make it more and more difficult for food companies to source non-GM ingredients, like soya, for food.

GM Freeze Co-ordinator Carrie Stebbings said:

'The public has made it clear that it wants foods produced from GM-fed animals to be labelled. Supermarkets have the ability to trace the majority of animal products back to the farms they have come from, so there is no reason for them not to specify non-GM animal feeds. The only explanation for their failure to do this appears to be cost – it seems for many companies shareholders come before the interests of their customers'.

Opinion poll surveys show that consumers have rejected GM, with the latest Eurobarometer poll concluding that 'overall Europeans think that GM food should not be encouraged. GM food is widely seen as not being useful, as morally unacceptable and as a risk for society.'

Meanwhile, the UK Government has launched a public consultation on GM and non-GM crop 'coexistence' in England, revealing plans to allow widespread GM contamination of conventional and organic crops.

Friends of the Earth is encouraging the public to contact their supermarkets and demand action to stop using GM animal feed: www.foe.co.uk/campaigns/real_food/press_for_change/gm_labelling/index.html
6 September 2006

⇨ The above information is reprinted with kind permission from Friends of the Earth. Visit www.foe.co.uk for more information.
© Friends of the Earth

International politics

Throughout Europe, public opinion is against GM crops and foods, yet the WTO ignores those concerns in favour of free trade

While the UK is likely to remain free of GM crops until 2009 at the earliest, other countries are only too keen to give genetic pollution a passport to contaminate the environment. The United States, Argentina, Canada are the main growers and exporters but developing nations such as China, Brazil, India and Thailand are also becoming increasingly involved in the global GM market.

The main GM crops being grown are maize soya, some of which still enters the UK as animal feed, and cotton. In terms of crops for human consumption, the EU has effectively been closed to imports of GM foods due to a piece of legislation requiring all products to be clearly labelled if they contain GM ingredients.

With consumer opinion in the EU still dead set against GM food, trying to sell the stuff would be commercial suicide. However, GM crops such as soya still enter the EU and the UK as animal feed, and the labelling regulations do not apply to meat and dairy products from animals raised on GM feed. Even though consumers had given a thumbs down to GM foods, we revealed how supermarkets like Sainsbury's were still selling milk from cattle fed on GM soya and maize.

But if we've all said no to GM food, who is trying to force it onto our plates? One of the main culprits is the US government, firmly in favour of a world swamped with GM food. It challenged the EU's anti-GM stance through the World Trade Organisation (WTO), a body whose sole purpose is to ensure a never-ending worldwide increase in trade and productivity, regardless of the social or environmental consequences.

The biotech companies that the US supports see the choice to know whether we are eating GM foods as a threat to their industry, and rightly so. But even though the WTO's ruling in 2006 stated that the EU had broken trade laws, it failed to give the US the leverage it needed to force GM foods into Europe as it determined that individual states were able to resist the introduction of GM foods if there was sufficient evidence of a danger to human health or the environment.

By reducing the arguments to ones of trade and economics, rather than biodiversity and the protection of the environment, the US and the biotechs are seeking to undermine the Biosafety Protocol. An addition to the internationally-agreed Convention on Biological Diversity, it is designed to regulate the international trade, handling and use of any GM organisms and at its heart is the precautionary principle. This principle allows countries to ban or restrict the movements of such organisms when there is a lack of scientific knowledge or consensus regarding their safety. *11 August 2006*

⇨ Information from Greenpeace. Visit www.greenpeace.org.uk for more information.

© Greenpeace

Potato research trials

Defra approves GM potato research trials

Defra has approved an application by the company BASF to undertake trials of a GM disease-resistant potato. The trials will take place on two sites in England, starting in 2007.

The BASF application has been evaluated by the independent expert group the Advisory Committee of Releases to the Environment (ACRE). It is satisfied that the trials will not result in any adverse effect on human health or the environment.

The GM potato developed by BASF is resistant to late potato blight. This can be a significant disease problem for UK potato growers, who normally combat it by applying chemical fungicides.

The purpose of the research trials is to test the effectiveness of the potato's resistance against UK strains of the disease. Similar trials are already under way in three other European countries.

Reflecting ACRE's advice, precautionary conditions have been attached to the statutory consent for the trials. These conditions will ensure that GM material does not persist at the trial sites. The harvested GM potatoes will not be used for food or animal feed.

Environment Minister Ian Pearson said:

'Our top priority on this issue remains protecting consumers and the environment, and a rigorous independent assessment has concluded that these trials do not give rise to any safety concerns.

'Based on the independent advice we have received, appropriate conditions have been specified for the conduct of the trials, and our GM Inspectorate will ensure that these are met. As the GM potatoes are being grown for research purposes they will not be used for food or animal feed.'

1 December 2006

⇨ The above information is reprinted with kind permission from Defra. Visit www.defra.gov.uk for more information.

© Crown copyright

Farmer quits GM trial after phone threats

**By Ian Sample,
Science Correspondent**

A Derbyshire farmer has pulled out of a GM crop trial due in the new year, citing fears for his personal safety.

The German plant science company BASF confirmed it was looking for a new site to conduct a trial of GM potatoes after the unnamed farmer in Draycott, Derbyshire, withdrew yesterday. He is believed to have received anonymous phone calls about his involvement in the trials.

The company was granted permission this month to plant GM potatoes at two single-hectare test sites in Derbyshire and at the National Institute for Agricultural Botany in Cambridge. The experimental potatoes are modified to resist late blight, the fungus that devastated Ireland's potato crop in the 1840s famine, and were expected to be planted in April. They would be incinerated after the trials.

BASF yesterday insisted the five-year trials would still go ahead, the first in Britain since the government's field-scale evaluations in 2003 to examine the environmental impact of herbicides used with some GM crops. Crop scientists were yesterday dismayed that tactics used by protesters during the height of the GM debacle of the 1990s seemed to have returned.

'What we find reprehensible are attempts by some groups to derail this sort of research by intimidation,' said Julian Little of the Agricultural Biotechnology Council, a GM crop industry group.

Derbyshire police last night said they were unable to comment on the situation.

The trial was approved by the Department for the Environment, Food and Rural Affairs after a green light from the government's advisory committee on releases to the environment (ACRE), which assesses all GM crop trials.

16 December 2006
© *Guardian Newspapers Limited 2007*

EU must wake up from 'GM food inertia'

Information from Foodnavigator.com

By Anthony Fletcher

When it comes to GM food, the EU needs to wake up from its political inertia, according to biotechnology pressure group EuropaBio.

The organisation, responding to what it sees as an opportunity being passed up, is holding a conference on 13 March 2007 in Lyon, France, to discuss the issue.

Panellists will include Dr Hans Kast, president and CEO of BASF Plant Science Holding, and Dr Bernward Gerthoff, chairman of the German Association of Biotech Industries-DIB.

'The proven benefits that green biotechnology can bring to farmers, the environment, consumers and society are already acknowledged and recognised by many at European level,' said the organisation.

'Despite a very stringent regulatory system for the assessment, approval and monitoring of agricultural biotech products put in place in Europe, there are still endless debates between opponents and advocates.

'Such debates result in a highly politicised European process for product authorisation that is very slow and in some instances prohibits the placing on the market of safe and beneficial products.'

EuropaBio claims that the consequence of this ambivalent position is the denial of freedom of choice for European farmers and consumers and negative influence on developing countries towards their adoption of biotech crops including those produced in their own countries to meet their own needs.

Indeed, within the European biotechnology sector, there is a real fear that the bloc is lagging behind the rest of the world in terms of access to agricultural biotechnology.

Marc Van Montagu, the president of the European Federation of Biotechnology, told journalists in Brussels recently that the technology, which has been oriented to helping developing countries, could also be of great benefit to European food production.

Montagu's comments follow the publication of new figures from the International Service for the Acquisition of Agri-biotech Applications (ISAAA).

The new statistics show that in 2006 the number of hectares globally cultivated with GM crops increased by 12 million hectares. Most of this growth came from countries such as China and India, while most EU farmers 'continue to be held back by a dysfunctional regulatory system and by disproportionate co-existence rules', according to Montagu.

The issue of GM approval within the EU is one of the most contentious in agriculture. The recent announcement that US authorities had traced amounts of unapproved genetically modified (GM) food in samples of rice prompted the EU to clamp down on all imports from the US.

The immediacy of this action illustrated the stringent controls the EU has in place to guard against unauthorised products entering the food chain, and also reflected consumer fears over the technology.

Nonetheless, in 2006, farmers cultivated approved biotech crops on 65,000 hectares in six European Member States (Portugal, Spain, Germany, France, Czech Republic and Slovakia). EuropaBio said that this would likely increase this year. *1 March 2007*

⇨ The above information is reprinted with kind permission from Foodnavigator.com. Visit www.foodnavigator.com for more information.

© *Foodnavigator.com*

Are EU GMO rules starving the poor?

Information from EurActiv

A debate organised by Friends of Europe, an EU policy think-tank, explored whether the EU's strict authorisation procedures on genetically modified food are preventing developing countries from investing in potentially life-saving technologies.

Background
On 20 February 2007, EU environment ministers voted against a Commission proposal to lift a ban imposed by Hungary on MON810 GM maize, which the country claims has harmful effects on European plants and animals.

This is the third time that member states have rejected Commission attempts to lift national bans on the growing of certain GM crops, despite assurances from the European Food Safety Authority's (EFSA) technical experts that they are safe.

EU ministers also failed to authorise the marketing of a genetically modified carnation – a sign that getting GM products approved in the EU has not become much easier since the EU's general moratorium – which effectively prevented any GMOs from being marketed in the EU for a five-year period – was lifted in 2003.

Large-scale GMO producers, such as the US, Argentina and Brazil, as well as large biotech companies including Monsanto, Sygenta and Bayer have been pushing for the EU to ease its authorisation procedure and let more GM crops in, resulting in a case at the World Trade Organisation (EurActiv 22/11/06).

Issues
A key argument put forward by GM producers is that GM technology could be the key to solving developing countries' hunger problem.

⇨ Does Europe have the right to systematically reject GMOs – even those that fulfil their own safety requirements?

⇨ Is Europe, through its stance on GMOs and strict authorisation procedures, stifling the development of a technology crucial to boosting food production and breaking the cycle of malnutrition and starvation in developing countries?

In a debate organised by think-tank Friends of Europe, green NGOs rejected this idea.

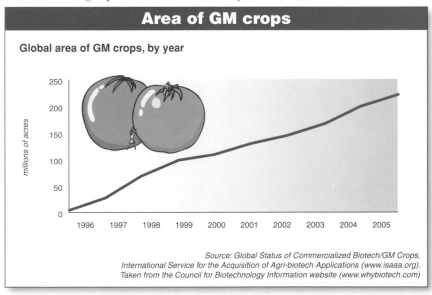

Area of GM crops

Global area of GM crops, by year

Source: Global Status of Commercialized Biotech/GM Crops, International Service for the Acquisition of Agri-biotech Applications (www.isaaa.org). Taken from the Council for Biotechnology Information website (www.whybiotech.com)

Positions

Danish Environment Minister Connie Hedegaard said that the EU should not dismiss all GMOs automatically, because the technology could help to solve developing countries' hunger problem.

'In a global world, the EU's actions impact on other countries,' she said, explaining that developing countries' inability to export to the EU discourages them from investing in and producing GMOs.

'Like it or not, GMOs are here to stay'

She believes that the scepticism in Europe about genetic engineering in agriculture stems from the fact that few GMOs 'have brought unquestionable benefits to the European table'. But she underlined the fact that the EU must assess each GMO on its own merits, because crops that can resist diseases and insects can be grown in the third world.

'Like it or not, GMOs are here to stay,' she said, adding that the EU has a special role to play in the debate because it can contribute to ensuring that GMOs are used in a safe and beneficial way for consumers by, for example, investing public research in this field.

Per Pindstrup-Andersen, Senior Research Fellow at the International Food Policy Research Institute (IFPRI), stressed: 'Not a single person has died or become sick because of GM foods.' Nevertheless, he agreed that more studies should be carried out on allergies, etc. 'The EU could have generated a lot of information on GMOs during the moratorium, but it simply sat on its hands,' he complained.

Although he conceded that Europeans have the right to know about the benefits and risks, he criticised the EU's dogmatism in refusing all GMOs.

'The debate in Europe is very one-sided,' he said, adding: 'If millions of farmers in India and China are willing to break laws to get genetically engineered food, there must be a reason.'

He underlined the importance of understanding the risk-benefit trade-off for developing countries, saying that for many the question is not 'Is genetic engineering the best solution?' but rather 'Is there any other solution?'

For the moment, he said, Europe is standing in the way of developing countries solving their own problems because of its straight-out rejection of GMOs. 'Developing countries are scared of losing their export market to Europe if they start cultivating GM crops,' he said.

But he agreed that Europe has an important role to play in encouraging the development of biosafety regulations, which are often very weak in developing countries.

Simon Barber, Director of External Relations, EuropaBio, the European Association for Bioindustries, said that the public had 'very limited knowledge' about GMOs and about agriculture in general. He accused green groups of spreading unfounded rumours, saying: 'After ten years of GM plants, what negative effects have ever been seen?'

He added: 'Many other plant-breeding technologies are just as scary and do not only produce benefits...To categorically say that the technology should not be used is not ethical.'

Furthermore, he said that imposing a ban on GMOs was not feasible anyway as 'the international trading system simply cannot segregate crops on a 100% basis'.

Fouad Hamdan, director of Friends of the Earth Europe (FoEE), believes that it is an exaggeration to say that GMOs can save developing countries, because there are only four types of GM crops: soy, maize, oilseed rape and cotton.

The majority of these crops are destined for feeding animals, not people, in rich countries.

Furthermore, he said, GM crops only benefit large farmers, not small ones who cannot afford expensive patented seeds. And, as for the environment, he said that the use of pesticides has actually increased in Europe following the introduction of GMOs.

He refuted the argument that NGOs were stirring up fear on false pretences, saying: 'I still believe that the benefits of GM food are almost nil...NGOs are working with independent scientific facts, not with biotech-industry funded research.'

Therefore, he concluded: 'The EU can with a lot of confidence tell developing countries to be cautious too. The organic market is the future.'

But a South African representative said: 'Most Africans don't have the luxury of choice of what to eat and what not to eat. If genetic engineering can bring some relief to this food insecurity, then let it be. And if it is too risky, then come up with another solution.'

23 February 2007

⇨ The above information is re-printed with kind permission from EurActiv. Visit www.euractiv.com for more information.

© *EurActiv*

Rice contaminated by GM

One-fifth of US rice contaminated with illegal GM strain

Up to one-fifth of rice entering the EU is contaminated with an illegal genetically modified (GM) strain from the US. Those are the findings of the European Commission's own investigation into EU rice imports, following the admission in August by the US government that untested strains of GM rice had entered the food chain.

If that wasn't alarming enough, our own research has shown this rice has made its way into products available in German supermarkets. Coming just one week after we revealed how Chinese products containing another illegal and untested GM rice variety were available on supermarket shelves in the UK and Europe, these results illustrate the inability of the GM industry to control its own technologies.

Out of 162 shipments of US long grain rice examined by the Commission, 33 tested positive for a strain of rice produced by agribusiness giant Bayer. The rice, LL601 as it's officially known, has been engineered to be resistant to Bayer's own herbicides but it has not been approved for human consumption

GREENPEACE

anywhere in the world. Currently, no varieties of GM rice have been approved for growing or consumption in the EU, although Bayer are trying to clear some of their other rice strains that have been approved in the US and Canada.

Illegal and untested

The rice was grown in the US in 2001 but only as a test crop and the effects on human health are unknown. Worrying, then, that it is now present on the shelves of Aldi Nord, a major German supermarket. Aldi Nord has since removed the affected products from its shelves but with Germany importing about 25 per cent of its rice from the US, the contamination could have spread much further.

There is already evidence to suggest this is the case. Testing by France and Sweden has shown that LL601 has washed up on their shores although this needs to be verified.

Meanwhile in the Netherlands, 20,000 tonnes of rice from the US have been detained in Rotterdam and out of 23 shipments tested so far, three came up positive. In the UK, the government is detaining all shipments of US long grain rice until they are confirmed contamination-free.

Speaking to Deutsche Welle, Ulrike Brendel, Greenpeace Germany GM campaigner, explained that the evidence is overwhelming. 'The fact we're finding it [in supermarkets] shows that the industry isn't capable of controlling genetically modified crops. We don't know what human health or environmental risks are involved. If we want to keep food sources free of genetically modified material, then we can't afford to plant GM crops.'

So what's to be done? Testing of rice and rice products on a worldwide scale would solve the problem in the short term but isn't a long-term solution. Jeremy Tager, GM rice campaigner with Greenpeace International, is quite clear about what is required.

'Once illegal GM crops are in the food chain, removing them takes enormous effort and cost,' he said. 'It is easier to prevent contamination in the first place and stop any plans to commercialise GM rice.'

Get active

You can have your say on the future of GM crops and food in the UK. The government is holding a consultation and you can submit your comments on their proposals. Friends of the Earth have produced an action pack with all you need to know to make your voice heard. so you don't need to be an expert on GM food, all you need is an opinion.
14 September 2006

⇨ The above information is reprinted with kind permission from Greenpeace. Visit www.greenpeace.org.uk for more information.

© *Greenpeace*

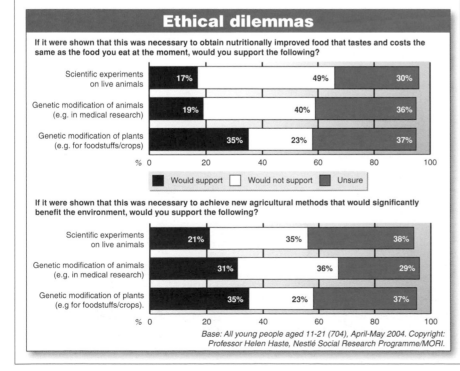

Ethical dilemmas

If it were shown that this was necessary to obtain nutritionally improved food that tastes and costs the same as the food you eat at the moment, would you support the following?

	Would support	Would not support	Unsure
Scientific experiments on live animals	17%	49%	30%
Genetic modification of animals (e.g. in medical research)	19%	40%	36%
Genetic modification of plants (e.g. for foodstuffs/crops)	35%	23%	37%

■ Would support   □ Would not support   ▨ Unsure

If it were shown that this was necessary to achieve new agricultural methods that would significantly benefit the environment, would you support the following?

	Would support	Would not support	Unsure
Scientific experiments on live animals	21%	35%	38%
Genetic modification of animals (e.g. in medical research)	31%	36%	29%
Genetic modification of plants (e.g for foodstuffs/crops).	35%	23%	37%

Base: All young people aged 11-21 (704), April-May 2004. Copyright: Professor Helen Haste, Nestlé Social Research Programme/MORI.

Legal challenge to FSA

Food Standards Agency taken to court over illegal GM rice

Illegal genetically modified rice was on sale in the UK more than two months after the Food Standards Agency (FSA) claimed it had been withdrawn from the market Friends of the Earth revealed today (Tuesday 20 February). The discovery of the GM rice, which may still be on sale in the UK, was revealed on the first day of a court case, brought by Friends of the Earth, against the FSA over its failure to take adequate steps to protect consumers from GM rice. The incident is the most significant GM food contamination episode to affect the UK.

> **'We have resorted to legal action to ensure that if another GM contamination incident happens, the FSA takes robust action to ensure that illegal GM ingredients are kept off our plates'**

The legal challenge centres on the FSA's failure to comply with an emergency EU law which instructed member states to remove unapproved GM rice from the market. The law was put in place after it was revealed, in August last year, that an experimental strain of GM rice (Bayer CropScience's LLRICE601), had contaminated commercial rice supplies in the US and been exported around the world. In the UK, contaminated rice was found in Tesco, Asda, Morrisons, Sainsbury's and Somerfield.

The FSA claims that by November 2006 there was no potentially contaminated US long grain rice on the market in the UK. However, Friends of the Earth has provided evidence that such rice was widely available for sale in London

Friends of the Earth

convenience stores in late January 2007. The FSA has confirmed that four of the rice packets purchased by Friends of the Earth from three different London stores were from batches that were contaminated with GM rice.

The FSA usually issues a food alert when food companies withdraw products to alert consumers and local authorities when there are problems with food, however, none were issued in this case.

Friends of the Earth's Head of Legal, Phil Michaels said:

'This experimental GM rice has not been licensed for human consumption in Europe. But the Food Standards Agency did not take adequate steps to prevent it being sold in the UK. Early on in this incident, the FSA decided to do nothing about contaminated rice products sitting on our shelves. We believe that the FSA has failed in its obligations to ensure that illegal GM rice was detected and removed from the market because it does not properly understand the approach to GM regulation that is required by European law.

'To justify its lack of action, the FSA relied on a consultant's model to estimate that contaminated rice would have been removed from the market after November. However, the fact that such rice was on sale more than two months after they claimed it had been removed shows their modelling was seriously flawed.'

Friends of the Earth's GM Campaigner, Clare Oxborrow, said:

'The FSA has failed consumers by attempting to wash its hands of any responsibility over this incident.

Instead of acting to make sure the public were not exposed to illegal GM rice, the Agency sat back and waited for contaminated products to be sold and eaten.

'We have resorted to legal action to ensure that if another GM contamination incident happens, the FSA takes robust action to ensure that illegal GM ingredients are kept off our plates. The FSA must also conduct routine tests of food imports from countries growing experimental GM crops to help prevent any more contamination incidents happening in the future.'

21 February 2007

⇨ The above information is reprinted with kind permission from Friends of the Earth. Visit www.foe.co.uk for more information.

Out of control?

Rice contamination proves GM is impossible to control

In August, the US notified the EU that it had detected trace amounts of unauthorised GM rice in its long-grain rice supplies intended for human consumption. Since then there have been over 107 incidents of contaminated rice found in 16 European countries. No one seems to know how the contamination occurred, or just how widespread it is. This incident highlights just how difficult it is to contain and monitor GM genes once they are in the environment.

Contamination found

Between 1998 and 2001 Bayer CropScience carried out trials in the US on a variety of herbicide-resistant rice it calls LL601. For reasons we do not know, Bayer decided not to continue with the process of commercial approval for LL601.

However, five years later, in August this year, the EC was notified that traces of the unapproved LL601 rice had contaminated long-grain rice grown in Arkansas, Missouri, Mississippi, Louisiana and Texas and may have been exported to Europe. No GM rice is authorised for sale anywhere in Europe, so any contaminated US rice on sale would be illegal.

The EU's immediate response was to only allow US long-grain rice certified to be free of LL601 to enter Europe. The rice had also not been fully assessed for safety and so no one can be certain whether it presents a health risk or not. Despite this, leaked minutes from a meeting with food companies showed that the Food Standards Agency (FSA), charged with protecting consumers and the food chain, told retailers that it did not expect them to remove rice already on their shelves that might be contaminated from sale, as it believed the rice to be safe to eat. The FSA advice to retailers ran contrary to requirements set down in the EU's emergency measures agreed in August, and also indicated to retailers that they could break the law by selling an illegal product.

This has led Friends of the Earth to start proceedings of a Judicial Review of the FSA handling of the contamination issue. The FSA subsequently revised its public advice that retailers should remove any contaminated rice already on sale, but it continues to state that LL601 is safe to eat.

Out of control

In the meantime, contaminated rice was found in products sold by Morrisons, Tesco, Sainsbury's and the Co-op. Friends of the Earth found LL601 in rice from Morrisons, who responded by assuring the public they had taken all contaminated rice off sale. However, a week later campaigners in the South West discovered rice from the same contaminated batch on sale in their local Morrisons which also proved to contain LL601.

Control also appeared to be a problem in European ports, as a shipment of US rice in Rotterdam, that had certificates from the US declaring it to be GM-free, turned out to contain LL601 when later tested. The EU then entered in to prolonged discussion with the US on a reliable certification system, but was unable to reach agreement and now requires all US rice shipments to be tested on arrival in EU ports before being distributed. Japan has gone even further, flying over samples from the US to be tested before allowing shipments to set sale from the US to Japan.

This disruption has caused serious financial problems for rice farmers in the US, who have launched at least 25 federal law suits against Bayer to try to recover their losses. However, some Americans seem to blame Europe's over-regulation for the loss of their markets, rather than accepting that unauthorised GM products are illegal in the EU and we have the right to uphold our own laws. A press release from the US Rice Federation refers to the EU actions as an 'unfortunate overreaction' that is 'denying EU consumers of wholesome American rice'.

In the US, Bayer's reaction to the crisis has been to apply for a fast-track deregulation of LL601, which would give it commercial approval to be sold there and limit the economic liability of Bayer having contaminated rice with an unapproved GM variety. Bayer has so far been unable or unwilling to reveal how contamination took place on such a large scale when LL601 had not been grown for five years.

What coexistence?

The incident occurred right in the middle of the Defra consultation on coexistence of GM, non-GM and organic crops and helps to highlight one of the many reasons why the concept of coexistence is a myth that will not protect our food supplies from GM contamination. It also highlights the difficulties we face in monitoring what is in our food that we might not be looking for.

Enforcement authorities across the EU were hampered by the lack of direct test materials for LL601 rice – Bayer only supplied this to a handful of labs. This is a common problem for all experimental GM crops being tested outside Europe or those awaiting EU commercial approval. A GM Freeze survey of local authorities in 2005 highlighted the serious lack of money devoted to testing for unwanted and illegal GM presence in imports.

The rice incident has shown that experimental lines are now contaminating the food chain, and no one would have been testing rice in Europe for GM because they would not have been expecting to find it. This raises the very significant question of what other foods are contaminated with GM traits that we are unaware of and are not looking for, and how do we ensure that we do know what we are eating?

In the US there are significant outdoor trials growing pharmaceuticals in food crops. This would be very difficult to detect in food, as the authorities here would not necessarily know what they were looking for when testing. It seems only a matter of time before we find GM pharma crops in our food supply.

GM Freeze believes that the only answer is to introduce international protocols whereby any country testing a GM trait in a crop that it exports must provide the countries it imports to with the templates for each experimental trait. Authorities in the EU should then regularly and routinely test imports for those experimental traits. The same should also apply to GM traits approved in a country exporting crops to countries where the GM traits have not been approved. Only then will we be in a position to accurately monitor and control GM contamination around the world.

November 2006

Update March 2007: Early in 2007, further GM contamination of long grain rice was discovered in a popular variety called Clearspring, leading the Californian Rice Commission to call for a moratorium on GM rice field trials. This contamination was with another of Bayer's GM varieties: LL62 GM rice. Further rice contamination with, as yet, unidentified GM trait has also been reported.

⇨ The above information is reprinted with kind permission from GM Freeze, and is taken from their newsletter *Thin Ice*. Please visit www.gmfreeze.org for more information.

© GM Freeze

Developing countries and GM crops

Information from the John Innes Centre

Background

⇨ Projections are that food production currently could feed 6.5 billion (assuming perfectly efficient distribution) but that we will need to feed 9 billion by 2050.

⇨ We therefore need to double food production by 2050 (because of changes in diet in rapidly modernising countries like China).

⇨ Global food deficit predicted for 2020 – a relatively short period in terms of scientific advance and technological application.

⇨ Pests and diseases cause up to 40% losses in many tropical crops.

⇨ Cultural and dietary preferences in under-producing countries often differ from crops and varieties in over-producing countries, e.g. white and yellow maize, and locally within a country, e.g. basmati rice in India.

⇨ Grain reserves are historically low in relation to demand.

⇨ There are no major new prime lands for grain production (except possibly the Ukraine). Therefore, increases must come almost entirely from technology.

⇨ The technology challenge is amplified by progressive loss of prime arable land by erosion, desertification and urbanisation and by global climate change and associated uncertainties in environmental concerns.

⇨ Small farmers in many developing countries benefited from the first Green Revolution but not in Sub-Saharan Africa.

⇨ Developing country farmers, breeders and agronomists are not averse to biotechnology – e.g., virus-free banana from tissue culture, marker-assisted breeding of rice.

⇨ International Research Institutes, national programmes and extension services provide networks for transferring technologies attuned to local needs.

What is needed – overview and specific examples

⇨ Conway's article on the small-scale woman farmer and her needs for stress-resilient and disease/pest-resistant crops, to secure food supply and generate income reliably, even in bad years.

⇨ No available genetic resistance to many pests and pathogens e.g. insect damage in cowpeas, virus infection of crops such as rice, maize, cassava or yams or fungus or nematode diseases of plantain.

⇨ Resilience to adverse environmental conditions – drought, salinity, poor soils, temperature fluctuations.

The experience so far

Bt cotton in China, South Africa and India leads the way.

⇨ Both Monsanto and Chinese transgenic lines in use in China, Monsanto lines in South Africa and Monsanto-derived lines in India.

⇨ Substantial decrease in the need for pesticide applications.

⇨ Slight increase in yield in China (c. 10%); increased yield in South Africa (up to 80%) and India.

⇨ Increased income to farmers because of reduced input costs and increased yields – in China the difference between modest profits and barely breaking even, in South Africa and India significant profits.

⇨ Less field work for women in South Africa.

⇨ Reduced health problems for farm workers from reduced exposure to pesticides.

⇨ Benefited small-scale farmers; in South Africa pump-priming the economy so that farmers can afford to buy new seed.

⇨ Possibility of developing money-generating associated commercial activities, e.g. local seed industry, product distribution system.

Examples of hurdles to uptake of GM crops

⇨ Release of sweet potato modified to give resistance to feathery mottle virus in Kenya was delayed for many years because of confusion over granting regulatory permission (defect in regulatory structure and impact of European concerns). Although proved effective very slow uptake because of infrastructural defects.

⇨ Attempts to produce transgenic papaya resistant to ringspot virus in South East Asia delayed because of need for characterisation of

the local strains of virus and inadequate research funding and infrastructure.

Examples of potential of currently available technologies

⇨ Immunisation through direct delivery of vaccines – e.g., via banana – efficacy, price, logistics.

⇨ Enhanced nutrition – golden rice for vitamin A, alleviation of iron deficiencies etc.

⇨ Resistance to important insect pests and diseases.

Photo: Arjen Lutgedorff

⇨ Enhanced agronomic performance for locally preferred varieties where conventional breeding has failed, e.g. semi-dwarf yield gene into Basmati rice.

Potential of emerging technologies and fundamental insights

⇨ The short-term major needs are for control of pests and diseases. Many technologies are available or emerging but need to be adapted to specific situations (crops/pests/diseases).

⇨ The longer-term needs include resistance to abiotic stresses such as drought and salinity so that poor unproductive land can be used and to adapt crops for post-harvest storage and processing so that the food flow from rural to urban areas is more effective.

Issues

⇨ Preservation of choice in open seed markets and ability of subsistence farmers to save seed if desired. Development of uptake pathways in many countries.

⇨ Renewed investment needed in R&D for poor people's ('orphan crops') crops – cowpea, millet, plantain, cassava, etc.

⇨ Renewed investment needed in research on developing country agronomic problems, e.g., the parasitic weed Stryga, environmental stress resilience.

⇨ Creative mechanisms to bridge technology gaps and develop crop R&D capacity in developing countries such as collaboration between industrialised and developing countries capitalising on their strengths – advanced technologies in industrialised countries and agronomic aspects in developing countries.

⇨ Education of developing country farmers and consumers to create a 'pull' environment for the technologies. This should reduce the 'hurdles' for the uptake of potentially beneficial technologies.

⇨ Development of robust bio-safety regulations and protocols in developing countries together with stewardship schemes – would involve enhancing extension services that, in turn, would increase farmer education.

⇨ Resolution of impact of GM controversy in Europe.

⇨ Resolution of IP and FTO issues.

References

1. Conway, G. (2000) *Environment* 42 (1): 9-18
2. Qaim, M. and Zilberman, D. (2003) *Science* 299, 900-902
3. Conway, G, Toenniessen G (2003) *Science* 299, 1187-1188
4. Toenniessen, G.H., O'Toole, J.C. and DeVries, J. (2003). Current Opinion in Plant Biology (in press).

Last updated 9 March 2007

⇨ The above information is reprinted with kind permission from the John Innes Centre. Visit www.jic.ac.uk for more information.

Plans to allow GM farming in secret 'are irresponsible'

By Charles Clover, Environment Editor

GM crops could be grown in secret under Government plans announced yesterday.

The move was denounced as 'irresponsible' by surveyors, who gave warning that it could blight land and property prices. Environmentalists said the proposed rules for 'co-existence' between genetically modified and other crops would lead to widespread contamination of the countryside.

Farmers would only have to notify neighbouring farmers if they were growing GM crops within a separation distance that could be as little as 35 metres (38yd) for GM oilseed rape.

The Government said it had decided against a public register for the growers of GM crops because of the cost and burden it would place on farmers

Farmers would be under no obligation to notify the owners of nearby gardens, allotments or beehives that they were growing GM varieties.

However, their neighbours could still find their produce contaminated by GM pollen, the effects of which can be measured over a kilometre (0.6 mile) away.

The Government said it had decided against a public register for the growers of GM crops because of the cost and burden it would place on farmers.

The rules it proposes will allow the contamination of neighbouring crops and honey up to the EU's legal threshold of 0.9 per cent of the crop – without any form of compensation. Environmentalists say that in Brazil non-GM crops have to conform to a threshold of 0.1 per cent, which farmers are able to do without difficulty.

The Government has made a number of proposals for compensating farmers who find they had been contaminated at more than 0.9 per cent by GM pollination. These do not go beyond the cost of the individual crop.

No GM crops suitable for UK conditions have been approved by the EU and it would take until 2009 for any to receive approval under its long-winded procedures. Ian Pearson, the environment minister, insisted yesterday that the Government was not for or against GM, and that proposals were 'not a green light for GM crops'.

The Royal Institution for Chartered Surveyors said it was 'disappointed' that the Government did not support the introduction of a GM land register.

Damian Cleghorn, an RICS spokesman, said: 'It is irresponsible of the Government not to introduce a land registry that would allow prospective purchasers of land and property to be warned about any possible issues relating to their transactions.

'A GM land register is in the public interest and it is the Government's responsibility to act in the public's interest'

'A GM land register is in the public interest and it is the Government's responsibility to act in the public's interest.'

Lord Melchett, policy director of the Soil Association, said: 'The Government's latest proposals are, in effect, denying all consumers, organic or non-organic, the right to choose non-GM food.'

21 July 2006

This crop revolution may succeed where GM failed

Gene splicing has been made obsolete by a cutting-edge technology that greatly accelerates classical plant breeding

By Jeremy Rifkin

For years, the life-science companies – Monsanto, Syngenta, Bayer, Pioneer etc. – have argued that genetically modified food is the next great scientific revolution in agriculture, and the only efficient and cheap way to feed a growing population in a shrinking world. Non-governmental organisations – including the Foundation on Economic Trends, of which I am president – have been cast as the villains in this agricultural drama, and often categorised as modern versions of the Luddites, accused of continually blocking scientific and technological progress because of our opposition to GM food.

Now, in an ironic twist, new cutting-edge technologies have made gene splicing and transgenic crops obsolete and a serious impediment to scientific progress. The new frontier is called genomics and the new agricultural technology is called marker-assisted selection (MAS). The new technology offers a sophisticated method to greatly accelerate classical breeding. A growing number of scientists believe MAS – which is already being introduced into the market – will eventually replace GM food. Moreover, environmental organisations that oppose GM crops are guardedly supportive of MAS technology.

Rapidly accumulating information about crop genomes is allowing scientists to identify genes associated with traits such as yield, and then scan crop relatives for the presence of those genes. Instead of using molecular splicing techniques to transfer a gene from an unrelated species into the genome of a food crop to increase yield, resist pests or improve nutrition, scientists are now using MAS to locate desired traits in other varieties or wild relatives of a particular food crop, then crossbreeding those plants with the existing commercial varieties to improve the crop. This greatly

> **New cutting-edge technologies have made gene splicing and transgenic crops obsolete and a serious impediment to scientific progress**

reduces the risk of environmental harm and potential adverse health effects associated with GM crops. Using MAS, researchers can upgrade classical breeding, and cut by 50% or more the time needed to develop new plant varieties by pinpointing appropriate plant partners at the gamete or seedling stage.

Using MAS, researchers in the Netherlands have developed a new lettuce variety resistant to an aphid that causes reduced and abnormal growth. Researchers at the US department of agriculture have used MAS to develop a strain of rice that is soft on the outside but remains firm on the inside after processing. Scientists in the UK and India have used MAS to develop pearl millet that is tolerant of drought and resistant to mildew. The crop was introduced into the market in India in 2005.

While MAS is emerging as a promising new agricultural technology with broad application, the limits of transgenic technology are becoming increasingly apparent. Most of the transgenic crops introduced into the fields express only two traits, resistance to pests and compatibility with herbicides, and rely on the expression of a single gene – hardly the sweeping agricultural revolution touted by the life-science companies at the beginning of the GM era.

There is still much work to be done in understanding the choreography, for example, between single genetic markers and complex genetic clusters and environmental factors, all of which interact to affect the development of the plant and produce desirable outcomes such as improved yield and drought resistance. Also, it should be noted that MAS is of value to the extent that it is used as part of a broader, agro-ecological approach to farming that integrates new crop introductions with a proper regard for all of the other environmental, economic and social factors that together determine the sustainability of farming.

The wrinkle is that the continued introduction of GM crops could contaminate existing plant varieties, making the new MAS technology more difficult to use. A landmark 2004 survey conducted by the Union

The new technology offers a sophisticated method to greatly accelerate classical breeding

of Concerned Scientists found that non-GM seeds from three of America's major agricultural crops – maize, soya beans and oilseed rape – were already 'pervasively contaminated with low levels of DNA sequences originating in genetically engineered varieties of these crops'.

Not surprisingly, MAS technology is being looked at with increasing interest within the European Union, where public opposition to GM food has remained resolute. In a recent speech, Stavros Dimas, the EU's environment commissioner, noted that 'MAS technology is attracting considerable attention' and said that the EU 'should not ignore the use of "upgraded" conventional varieties as an alternative to GM crops'.

As MAS becomes cheaper and easier to use, and as knowledge in genomics becomes more easily available over the next decade, plant breeders around the world will be able to exchange information about best practices and democratise the technology. Already plant breeders are talking about 'open source' genomics, envisioning the sharing of genes. The struggle between a younger generation of sustainable-agriculture enthusiasts anxious to share genetic information and entrenched company scientists determined to maintain control over the world's seed stocks through patent protection is likely to be hard-fought, especially in the developing world.

If properly used as part of a much larger systemic and holistic approach to sustainable agricultural development, MAS technology could be the right technology at the right time in history.

⇨ Jeremy Rifkin is the author of *The Biotech Century*. Article first appeared in the *Guardian*, 26 October 2006.

© *Jeremy Rifkin 2006*

There's no chance that this technology will replace GM

Genetic modification will remain a vital tool in the global production of crops, says Tony Combes

I seemed to have heard Jeremy Rifkin's advocacy of marker assisted selection (MAS) plant breeding – 'new', 'cutting edge' – somewhere before ('This crop revolution may succeed where GM failed', 26 October). I had. In 2001, Rifkin extolled MAS in the *New York Times*: 'I think that's where the future is,' he said.

In order to ensure future agricultural sustainability, plant breeders and scientists need access to a toolbox of technologies

In order to ensure future agricultural sustainability, plant breeders and scientists need access to a toolbox of technologies. Traditional breeding by cross-pollination, MAS and GM technology are like having three grades of toolbox. The first is tried and trusted, but the tools are limited and precision is difficult. MAS allows one to do a bit more, focusing better on some more specific objectives. But the way to do things exactly is to know just what you want to do and to have a set of precision tools for doing it. GM technology is well on its way to achieving that, transferring only the benefits farmers want in order to improve their crops, without the limits of traditional plant breeding.

In the words of Mike Gale, an emeritus cereal geneticist at the John Innes Centre in Norwich: 'If we are going to produce enough food locally to help feed the world, plant breeders need every tool in the toolbox.'

Rifkin is letting his hatred of the use of biotechnology in agriculture get ahead of his better judgement when suggesting MAS could be a 'replacement' for GM techniques. The two technologies are very different.

Fuelling repetitive media scare stories, while attempting to marginalise those farmers who choose GM crop benefits (and why else would GM harvests repeatedly increase?) is the real failure here

---

For example, Rifkin claims that MAS works by locating 'desired traits in other varieties or wild relatives of a particular food crop, then crossbreeding those plants with the existing commercial varieties to improve the crop'. That is a very blunt tool; it can take years to identify the correct markers. In comparison, gene splicing inserts a beneficial gene into a plant, thereby imparting a specific property. That might, for example, be herbicide tolerance, reducing the amount of herbicide the farmer needs to spray – as the United States Department of Agriculture reported is happening in America. Or it might be resistance to insect attack, saving countless applications of old-fashioned insecticides especially in resource-poor countries.

And it now includes drought tolerance, enabling plants to grow in the semi-arid conditions which the Stern report last week identified as one of the most serious challenges of global warming in the developing world. Crops are already being field tested, but will activists permit African farmers to benefit in time?

All these advances are bringing hugely important agro-envir-onmental benefits, but we know little about what MAS can deliver. Not surprisingly, even European farmers can now choose the benefits of GM corn/maize – those in Spain have done so continuously since 1998.

The Agricultural Biotechnology Council's role is to promote understanding through open debate on GM crops and technologies, hence we deplore the misinformation from all NGOs' anti-GM campaigns. Fuelling repetitive media scare stories, while attempting to marginalise those farmers who choose GM crop benefits (and why else would GM harvests repeatedly increase?) is the real failure here.

⇨ Tony Combes is deputy chairman of the Agricultural Biotechnology Council, and director of corporate affairs for Monsanto UK (www.abcinformation.org).
7 November 2006
© *Guardian Newspapers Limited 2007*

Tories rally against Frankenstein foods

The Tories have joined forces with organic farmers to warn that government plans to allow GM farming will contaminate their crops and destroy their industry.

The Government has produced draft proposals allowing genetically-modified crops to be grown commercially in this country.

But critics warn that the pollen from these farms will inevitably spread. This will effectively destroy the right of consumers to choose not to eat food tainted with GM and do lasting damage to the organic food industry.

Under the plans, the GM pollen pollution of conventional and organic crops or honey will be legalised up to a threshold of 0.9 per cent.

Any crop contaminated up to this level would not need to be labelled, effectively leaving consumers in the dark about what they are eating.

Farmers whose crops suffer higher levels of contamination will be entitled to little, if any, compensation.

An alliance of 74 organic businesses, the Food & Drink Federation and the Soil Association will meet at the Commons today to challenge the GM plan.

Tory food and farming spokesman Peter Ainsworth MP said: 'This is an issue about consumer choice. Scientists tell us they can trace GM content to 0.1 per cent and consequently food labelling should reflect this.

'With the concerns surrounding GM food we believe that the public should be told the GM content so they can make an informed choice about what they buy.'

The system for assessing the safety of pesticides is to be challenged in the courts.

Campaigner Georgina Downs argues that the Government's claims that crop-spraying is safe are not backed up by proper research and evidence.

And a High Court judge has decided her concerns should be investigated through a judicial review.

Miss Downs said: 'There has never been any risk assessment for the long-term exposure for those who live, work or go to school near pesticide-sprayed fields.'

⇨ This article first appeared in the *Daily Mail*, 4 February 2007.
© *2007 Associated Newspapers Ltd*

KEY FACTS

⇨ Contemporary GM techniques are based on scientific discoveries made in the 1950s. Research into molecular biology and genetics in the 1970s resulted in the first GM plants being bred during the early 1980s. The first commercial crops were grown on a large scale in 1996. (page 2)

⇨ Overall public attitudes towards GM foods are still negative with 47 per cent of people feeling that GM crops should not be grown for commercial use and with 29 per cent of people still undecided on the issue. (page 3)

⇨ There are currently no GM crops being grown in the UK. (page 5)

⇨ Under EU legislation each proposed release of a GM product is subject to a detailed risk assessment which involves careful scrutiny by independent scientists. (page 6)

⇨ The USA has the highest proportion of GM crops being grown, with an area of 54.6 million hectares dedicated to GM crops. This is 36.6 million hectares more than the next highest GM-producing country, Argentina, at 18 million hectares. (page 8)

⇨ People disagree about whether GM crops will help solve the world food crisis, or whether GM is just a 'technical fix' for a much more fundamental problem. (page 10)

⇨ EC figures show that 23% of Europeans would definitely buy GM foods if they could be shown to be healthier. 33% said that they probably would. (page 17)

⇨ 50% of Europeans aged under 25 said that they would buy GM foods if they were cheaper, compared to only 32% in the 46 to 65 age group. (page 17)

⇨ The first report into the extent to which genetically engineered (GE) organisms have leaked into the environment reveals a disturbing picture of widespread contamination, illegal planting and negative agricultural side effects, says GeneWatch. (page 19)

⇨ A flock of designer hens, genetically modified with human genes to lay eggs capable of producing drugs that fight cancer and other life-threatening diseases, has been created by British scientists. (page 20)

⇨ Like other GM technologies, pharming is not without its risks. Pressure groups such as Friends of the Earth fear that if food crops such as maize or tomatoes are adopted to grow drugs in some regions, there is a risk of their contaminating maize or tomato crops elsewhere that are intended for consumption. (page 21)

⇨ Tight controls are needed to ensure that GM crops do not contaminate natural plants. Growing in air-tight greenhouses prevents pollen escaping, but an alternative is to grow GM crops that have no relatives they can pollinate. Scientists are also working on infertile GM crops that do not flower. (page 22)

⇨ Materials from GM crops are used in animal feed in the UK, and are subject to a safety assessment as part of their authorisation. On the basis of these assessments, there is no reason to suppose that GM feed presents any more risk to farmed livestock than conventional feed. (page 25)

⇨ The global area of GM crops for 2005 was 90 million hectares in 21 countries, up from 81 million hectares in 17 countries in 2004 and 67.7 million hectares in 18 countries in 2003. (page 25)

⇨ A new survey by Friends of the Earth reveals that most animal products sold in supermarkets, including milk, cheese and meat, come from animals fed on GM feeds. But consumers are not aware of what they are buying, with five out of 10 supermarkets failing to tell customers when food comes from animals fed on genetically modified feed. (page 26)

⇨ The main GM crops being grown are maize, soya and cotton. (page 27)

⇨ Defra has approved an application by the company BASF to undertake trials of a GM disease-resistant potato. The trials will take place on two sites in England, starting in 2007. (page 27)

⇨ A key argument put forward by GM producers is that GM technology could be the key to solving developing countries' hunger problem. (page 29)

⇨ Up to one-fifth of rice entering the EU is contaminated with an illegal genetically modified (GM) strain from the US. Those are the findings of the European Commission's investigation into EU rice imports. (page 31)

⇨ In a survey by the Nestlé Social Research Programme, 35% of young people aged 11 to 21 said they would support genetic modification of plants if it was necessary to obtain nutritionally improved food that tastes and costs the same as the food they eat at the moment. (page 31)

GLOSSARY

Biodiversity
This term refers to the number and variety of different species among living organisms. There are concerns that the crossing of genes between different species through genetic modification would result in decreased biodiversity, which is not desirable.

Biofuel
A gaseous, liquid or solid fuel derived from a biological source, e.g. ethanol, rapeseed oil. Some scientists claim that GM would be a useful tool in the quest to produce biofuels which would be beneficial for the environment.

Biotechnology
Although this term is sometimes used interchangeably with genetic modification, it actually has a much wider scope. It refers to any use of a biological system or living organism to engineer or manufacture a product or substance.

Cross-contamination
Sometimes, genetically modified material can be passed unintentionally between plant crops: this is called cross-contamination. It can result in the presence of GM substances in what is supposed to be non-GM food, and is therefore one concern raised by consumers who want to be able to make informed choices about whether or not to eat GM food.

Farm-scale evaluation (FSE) trials
These were conducted in the UK in 2003 to measure the impact on farmland wildlife of the herbicide use associated with four GM herbicide-tolerant crops, as compared to the herbicide use associated with the equivalent conventional crops.

'Frankenstein foods'
A term coined by parts of the British media to describe foods produced using biotechnology.

Genes
A gene is an instruction and each of our cells contains tens of thousands of these instructions. In humans, these instructions work together to determine everything from our eye colour to our risk of heart disease. The reasons we all have slightly different characteristics is that before we are born our parents' genes get shuffled about at random. The same principles apply to other animals and plants.

Genetically modified organisms (GMO)
A genetically modified organism is one in which the genetic material has been altered by the direct introduction of DNA.

Genetic modification
May also be called modern biotechnology, gene technology, recombinant DNA technology or genetic engineering. Scientists are able to modify genes in order to produce different characteristics in an organism than it would have produced naturally. GM techniques allow specific genes to be transferred from one organism to another, including between non-related species. This technology might be used, for example, to produce plants which are more resistant to pesticides, which have a higher nutritional value, or which produce a greater crop yield. Those in favour of GM say that this could bring real benefits to food producers and consumers. Those against GM feel it is risky as scientists do not have the knowledge to 'play God' with the food we eat.

Genomics
An organism's genome can be defined as all its genetic material packed together. Genomics is the science of genomes – more specifically, their sequencing, mapping, analysis, study and manipulation.

Marker assisted selection (MAS)
A new agricultural technology which builds on traditional selective breeding practices. Some scientists believe it could provide an alternative to GM techniques in producing food crops.

'Pharming'
Pharmaceutical, or 'pharmed', crops are crops that have been genetically modified to produce medicines or for other pharmaceutical purposes.

Selective breeding
Human beings have been modifying the genes of biological organisms for centuries through selective breeding: choosing individual plants and animals with particular traits, like fast growth rates or good seed production, and breeding them with others to produce the most desirable combination of characteristics. However, unlike genetic modification, this can happen only within closely-related species.

'Terminator' seeds
The official name for these is Genetic Use Restriction Technologies (GURTs). Plants bred with this type of GURT have been genetically modified to produce sterile seeds, preventing a farmer from using them for future planting.

INDEX

Additional Resources

Other Issues titles

If you are interested in researching further some of the issues raised in *A Genetically Modified Future?*, you may like to read the following titles in the **Issues** series:

⇨ Vol. 134 *Customers and Consumerism* (ISBN 978 1 86168 386 1)

⇨ Vol. 119 *Transport Trends* (ISBN 978 1 86168 352 6)

⇨ Vol. 110 *Poverty* (ISBN 978 1 86168 343 4)

⇨ Vol. 98 *The Globalisation Issue* (ISBN 978 1 86168 312 0)

⇨ Vol. 95 *The Climate Crisis* (ISBN 978 1 86168 303 8)

⇨ Vol. 90 *Human and Animal Cloning* (ISBN 978 1 86168 291 8)

⇨ Vol. 88 *Food and Nutrition* (ISBN 978 1 86168 289 5)

For more information about these titles, visit our website at www.independence.co.uk/publicationslist

Useful organisations

You may find the websites of the following organisations useful for further research:

⇨ Ban Terminator: www.banterminator.org

⇨ CropGen: www.cropgen.org

⇨ Defra: www.defra.gov.uk

⇨ Food Standards Agency: www.food.gov.uk

⇨ Friends of the Earth: www.foe.co.uk

⇨ GeneWatch UK: www.genewatch.org

⇨ GM Freeze: www.gmfreeze.org

⇨ GM Nation?: www.gmnation.org

⇨ Greenpeace: www.greenpeace.org.uk

⇨ John Innes Centre: www.jic.ac.uk

⇨ Natural Environment Research Council: www.nerc.ac.uk

ACKNOWLEDGEMENTS

The publisher is grateful for permission to reproduce the following material.

While every care has been taken to trace and acknowledge copyright, the publisher tenders its apology for any accidental infringement or where copyright has proved untraceable. The publisher would be pleased to come to a suitable arrangement in any such case with the rightful owner.

Chapter One: GM Trends

What is GM?, © CropGen, *Genomics in the UK*, © ESRC, *Glossary*, © GM Nation?, GM (*Genetic Modification*), © Crown copyright is reproduced with the permission of Her Majesty's Stationery Office, *Why GM?*, © GM Nation?, GM *labelling*, © Crown copyright is reproduced with the permission of Her Majesty's Stationery Office, *Genetic modification and the environment*, © NERC, *Scientists to create healthier tomatoes*, © National Science Foundation, *Biotechnology: growing in popularity*, © European Commission, *UN upholds moratorium on terminator seed technology*, © Ban Terminator, GM *contamination*, © GeneWatch UK, GM *drug crops*, © Friends of the Earth, *Modified hens lay eggs to help beat cancer*, © Telegraph Group Ltd, GM *tobacco could save lives*, © Guardian Newspapers Ltd.

Chapter Two: The GM Debate

What are the ethics?, © GM Nation?, *What's the problem?*, © Greenpeace, GM *material in animal feed*, © Crown copyright is reproduced with the permission of Her Majesty's Stationery Office, *Supermarkets supporting GM through the back door*, © Friends of the Earth, *International politics*, © Greenpeace, *Potato research trials*, © Crown copyright is reproduced with the permission of Her Majesty's Stationery Office, *Farmer quits GM trial after phone threat*, © Guardian Newspapers Ltd, *EU must wake up from 'GM food inertia'*, © Foodnavigator.com, *Are EU GMO rules starving the poor?*, © EurActiv, *Rice contaminated by GM*, © Greenpeace, *Ethical dilemmas*, © Professor Helen Haste, Nestlé Social Research Programme, *Legal challenge to FSA*, © Friends of the Earth, *Out of control?*, © GM Freeze, *Developing countries and GM crops*, © John Innes Centre, *Plans to allow GM farming in secret 'are irresponsible'*, © Telegraph Group Ltd, *This crop revolution may succeed where GM failed*, © Jeremy Rifkin, *There's no chance this technology will replace GM*, © Guardian Newspapers Ltd, *Tories rally against Frankenstein foods*, © Associated Newspapers Ltd.

Illustrations

Pages 1, 19, 32: Angelo Madrid; pages 2, 30: Bev Aisbett; pages 4, 24, 38: Simon Kneebone; pages 7, 22, 33: Don Hatcher.

Photographs

Page 6: Herbert Berends; page 9: James Wilsher; page 16: Grzegorz Rejniak; page 26: Marja Flick-Buijs; page 28: Chris Johnson; page 35: Arjen Lutgedorff; page 37: Sanja Gjenero; page 39: Lionel Titu.

And with thanks to the team: Mary Chapman, Sandra Dennis and Jan Haskell.

Lisa Firth
Cambridge
April, 2007

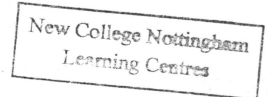